Science and the Supernatural

Science and the Supernatural

AN INVESTIGATION OF
PARANORMAL PHENOMENA
INCLUDING PSYCHIC HEALING,
CLAIRVOYANCE, TELEPATHY AND
PRECOGNITION BY A
DISTINGUISHED PHYSICIST AND
MATHEMATICIAN

John Taylor

E. P. DUTTON • NEW YORK

For information contact: E. P. Dutton, 2 Park Avenue,
New York, N.Y. 10016

Library of Congress Cataloging in Publication Data
Taylor, John Gerald
Science and the supernatural.
Bibliography: p.
1. Psychical research. I. Title.
BF1031.T32   1980   133.8   79-22302

ISBN: 0-525-19790-7

Published simultaneously in Canada by Clarke,
Irwin & Company Limited, Toronto and Vancouver

Designed by Barbara Huntley

10 9 8 7 6 5 4 3 2 1

First Edition

Contents

Preface   vii

1 Introduction   1

2 The Inexplicable Paranormal   10

3 The Framework of Science   20

4 Psychic Healing   31

5 Clairvoyance   46

6 Telepathy   57

7 Precognition   73

8 Psychokinesis   85

9 The Framework of the Paranormal: Fraud and Mischief
   109

10 The Framework of the Paranormal: Cues, Fantasy and
   Memory   125

11 The Framework of the Paranormal: Coincidence, Credulity
   and the Fear of Death   147

12 Unresolved Problems   162

Further Reading   171

Index   175

Preface

THIS IS A BOOK ABOUT THE PARANORMAL, THE SUPERNATURAL OR
ESP—about things that go bump in the night, or people who have
apparently lived earlier lives or have regular conversations with
the dead. It is about the sixth sense, that faculty whose possession
would enable one to take over the world if one possessed it to a
high enough degree; and whose possession, even to a lesser
extent, could smooth one's way through life.

An avalanche of books on the supernatural has been churned
out in the last decade. The human race appears to have a great
thirst for the unknown, the bizarre, the spine-chilling. All of these
books feed on this appetite with ever more blood-curdling tales,
supposedly true. This is not another book in that vein. It is
instead an account of one scientist's attempt to discover if the
supernatural really does exist. What is the truth behind the accounts? Do things happen exactly as they were reported about
the supernatural? Or is there only human credulity, fallibility,
fear and greed at work, causing natural phenomena to be completely misinterpreted and distorted? In other words, is the supernatural just a pack of lies?

I started my investigations into ESP because I thought there
could be something in it. There seemed too much evidence
brought forward by too many reliable people for it all to disap-

pear. Yet as my investigation proceeded that is exactly what happened. Every supernatural phenomenon I investigated crumbled to nothing before my gaze. That is why I have given this book the title it has. It could as well be titled "The Natural Supernatural" or alternatively "The Lying Supernatural." The first title would indicate that nature became supreme by the end of my investigation; the second, that error and deceit became more and more relevant for me in understanding the supernatural as my work proceeded.

My present attitude seems to be a complete about-face. My public persona over the last five years has been that of the scientist-manqué who has dared to investigate the supernatural; an area into which scientists hardly ever dare to tread. I even began by espousing the supernatural, believing abnormal radio-wave emissions by what has become known as "sensitives" might be the cause of paranormal phenomena. To disarm the obvious criticism let me say quite bluntly that I am not trying to regain any lost scientific "respectability" by my new position. Only through constant investigations of many sensitives over a number of years have my earlier positive views on the paranormal been required to be so radically revised.

Part of this program of research involved attempting to relate the paranormal to science itself. I had to try and answer questions: Is there a fifth force of nature only observed through paranormal phenomena? Does science have any leeway in order to explain some, if not all, of the various supernatural phenomena or are the laws of science contradicted by their existence? The answers to these questions are given in the book. In summary these answers indicate that there is a clear contradiction between science and most supernatural phenomena. The remaining phenomena can be given a definite, or at least highly probable, scientific explanation. From my own investigations the evidence for those supernatural phenomena contradicting science also seems rather doubtful, as I indicated earlier. Even if a fifth force did exist to explain such dubious cases it would still have to disagree with basic scientific principles. I could not therefore believe in both science and the supernatural simultaneously. On weighing science and the supernatural the scales seem to be

so heavily biased for science that until further evidence is produced I no longer believe in the supernatural.

Many have helped in the progress of this work, but above all
I would like to thank R. Ibrahim, E. Balanovski and A. Smith for
their direct help and to many colleagues at King's College and
elsewhere for their patience.

Science and the Supernatural

1

Introduction

THERE IS A WHOLE RANGE OF PHENOMENA THAT SOME OF US MAY experience at one time or another, and which runs counter to common sense. These occurrences are termed "psychic" if they are regarded as caused by the soul, but may be given the more general adjective "supernatural," as being above or outside the natural run of things. They are also called "paranormal," so as to remove the necessarily immaterial aspect associated with the word "psychic"; this places them alongside the normal, only awaiting explanation to be made respectable. The prefix "para" is then used to denote the extension of that branch of investigation, such as psychology or physics, to paranormal phenomena; hence "parapsychology" and "paraphysics."

All of these phenomena presently involve unknown powers of human beings beyond the more limited five senses to which we are so accustomed. The range of such supernatural happenings is indeed wide, including survival after death, the ability to look into the future (precognition), the power of the mind to move matter without having to touch it (psychokinesis), the direct communication of one mind with another (telepathy), the viewing of distant scenes (clairvoyance or distant viewing), the sensing of underground water or other objects (water divining or dowsing), the diagnosis and healing of the sick outside Western medicine

1

(faith or psychic healing), communication with the dead (direct voice mediumship) and numerous other powers. It extends even into the world of the occult, where the conjuring of discarnate entities is supposed to be achieved by certain methods.

Belief in the existence of psychic phenomena is both widespread and deep. For example a traveling exhibition of psychic phenomena organized in 1976 by the Smithsonian Institution in Washington, D.C., caused an enormous amount of interest. Tens of thousands of people flocked to see it. When the exhibit stopped at one place, the Morehead Planetarium in Chapel Hill, North Carolina, thirty thousand people saw it; as the planetarium director said, "It was tremendously successful."

No wonder that there is such concern when roughly four out of five Americans have psychic experiences ranging from telepathy to communications with the dead. A psychiatrist in Montclair, New Jersey, has recorded more than four thousand psychic experiences among his patients alone. A Brooklyn psychologist has been quoted as saying that America "has been a nation of mystics from the word go. Most of my patients have had a paranormal experience, and many friends report having them—including communications with the dead, clairvoyance, precognition, and déjà vu (the feeling of having been some place before). Based on my experience I'd say about 80 percent of the population has at least one mystical event." A Philadelphia psychiatrist is quoted as saying, "The vast majority of my patients have had a mystical experience, and there's no reason to think the rest of the population is any different."

An increasing acceptance of the possibility of ESP has also occurred among American university professors. A poll answered by 1,188 professors at major universities and colleges in 1977 (49 percent of those polled) indicated that 16 percent believed ESP was an established fact and another 49 percent considered it to be a likely possibility. In other words, 65 percent were favorable to ESP; only 4 percent believed it to be an impossibility. A similar study twenty-five years ago found only 17 percent to be favorable to ESP and 10 percent believed it to be impossible.

In Iceland, a university survey showed that 64 percent of the population have had a psychic experience. These results and those of a similar nature from other countries give evidence for an impressive level of concern about the paranormal throughout the world.

This interest has been reflected by concern at the government level in various countries. It is reported that at least five-figure sums are being spent by the U.S. Defense Department to support telepathy tests at Stanford Research Institute, while the United States Army Missile Research and Development Command is funding research to the tune of thirty thousand dollars into a particular technique called Kirlian photography, supposedly able to photograph the human aura.

Government interest in the phenomenon of "unidentified flying objects" (UFOs) has increased. Many of these have been observed by individuals in various parts of the world in the last few decades. An American government-sponsored investigation under the scientist Edward Condon failed to find any significant evidence for UFOs in the sixties, but a new UFO probe has been launched in France, backed by the French Defense Ministry. As part of this study, French universities and space research laboratories will devote time and personnel to investigate the validity of UFO reports. President Jimmy Carter's science adviser, Dr. Frank Press, has also asked American space officials to do a feasibility study in the U.S.A. to see if a major UFO study should be launched there. At first sight, UFO phenomena do not seem to be closely linked to the existence of a human sixth sense. Yet there are claims that contact with alien spacecraft—the UFOs —can best be made by such a human power, and in this way the UFO story becomes strongly linked with that of the supernatural.

There are various reports, some by emigré Russian scientists, that the Russians are seeking military uses for ESP results. This has led to fears in the United States that such research might uncover some effective method of "thought control" or "psychological warfare." The American Menninger Foundation, for example, has strongly suggested that U.S. scientists learn all they can about ESP before the Russians get the upper hand.

As is to be expected, there has developed rivalry between America and Russia over such investigations. One American psychologist commented, "I have heard from people involved in

Kirlian photography that the Defense Department wasn't interested in it until learning how advanced the Russians were."

In June 1977 the Russian authorities detained an American reporter, Robert Toth of the *Los Angeles Times,* and charged him with gathering secret information. After lengthy questioning, Toth was expelled from Russia. On his release, he said that a Russian scientist had handed him an article he had written to prove the legitimacy of parapsychology when the Russian police arrested them both. The Russian security police apparently accused Toth of having received "state secrets" about parapsychology.

It is one thing to believe in the supernatural; it is altogether another to prove that it exists. The more strongly a belief is held, the harder it is to see evidence for its disproof. This bias, and even sometimes clear gullibility, on the part of so many of the believers has made the sceptics even more sceptical. But the latter can be as irrational in their nonacceptance of any and all of the supernatural as the believer on his side can be in swallowing it hook, line and sinker.

Alongside the recent growth in interest in the paranormal has been a comparable increase in vociferous scepticism. One particular group has recently been formed in the U.S.A. under the impressive title, "The Committee for the Scientific Investigation of Claims of the Paranormal." One interested reporter termed this "a high-sounding title for a group which contains most of the rump of the hard-core U.S. sceptics. It is an anti-occult cult, and quite as irrational as many of those which it denounces, dedicated not to the investigation but to the demolition of parapsychology. And it likes to link parapsychology to little green men to promote guilt by association."

There is indeed good reason for healthy scepticism when faced with the supernatural, but it must be tinged with a certain degree of open-mindedness. This is not always easy to achieve if the facts that are staring you in the face will totally destroy your understanding of the world. This was the position I found myself in when I observed the Israeli psychic/entertainer Uri Geller for the first time, and heard of all the similar events that had happened in the homes of numerous viewers or listeners to media programs in which Geller had taken part. I lived with that sensation for several days before I could pull myself together and try to make sense of the phenomenon.

It all started for me on the evening of Friday, November 23, 1973. I had been invited by the BBC to act as a scientific hatchet man on Uri Geller during his appearance on David Dimbleby's "Talk-In." This was transmitted live, and since it was being screened just after the "Miss World" competition, it had many millions of viewers. I myself felt keyed up, both because of the live transmission (where mistakes cannot be remedied) and also because of Geller's supposed powers. I was almost sure he would not be able to do what he claimed—bend spoons or forks, start watches and guess pictures sealed in envelopes. Yet there was a small doubt in my mind. What if he *could* do these things?

On being introduced to Geller I found him to be an engaging and friendly young man, without a hint of the deviousness supposedly associated with him. Almost immediately after the program started a tray of assorted pieces of cutlery was brought on by a BBC assistant. Geller selected four big, solid forks and handed one to each of us on the program (myself, Dimbleby and Lyall Watson, the biologist and author), keeping one for himself. We sat and held them for a minute or so until suddenly Geller claimed that Watson's fork was bending. Indeed it was, and I watched in utter astonishment as Geller gently stroked the neck of the fork to cause it to break into two pieces. I could have sworn that the fork had not been forcefully bent back and forth during the program and I am certain that the many viewers at their television sets would have supported me in this.

Geller then demonstrated his telepathic powers by attempting to reproduce the picture of a yacht (in silhouette) which had been drawn before the program by a young BBC employee and sealed in an envelope. Again I had not expected much success, but Geller was absolutely accurate. I was once more astonished by this prowess; it just could not happen!

The final item had to do with starting or stopping watches. A number of broken watches had been gathered and Geller held his clenched fists over them, willing them to start again. As amazingly as before, he proceeded to make one of them start and, even more amazingly, cause the second hand of another to become buckled under the watch glass.

There was an enormous uproar in the studio as the "Talk-In" program ended. The invited audience milled about, talking at the tops of their voices about what they had seen demonstrated by Geller, some saying how it was impossible (the complete sceptics) and others that there were even more exciting things that could be achieved by such powers (the believers).

What was more interesting for me was the response of the viewing audience. The BBC telephone lines were completely swamped by callers saying that they had experienced in their own homes the various phenomena Geller had demonstrated. Knives, forks, spoons and even pots and pans had apparently become bent, in some cases without any direct contact with them. Watches (and even electric clocks), hitherto unusable, had started on command. Many had guessed the sealed picture of the yacht.

As I heard this, I was even more astonished. How could all these hundreds of people have suddenly discovered such fantastic powers? Why had they lain dormant until Geller's appearance on television triggered them off? More importantly, I felt it unlikely that what Geller had achieved was by fraud, unless some gigantic hoax was being perpetrated. But this made my faith in science even more at risk, for I just could not see how there could be even a glimmer of a scientific explanation for these phenomena. The scientific framework with which I viewed the world up till then was crumbling about my ears.

After the initial shock wore off, I began to realize that I should attempt to find out in more detail what was actually happening to the pieces of cutlery as they bent and how the requisite energy was being transmitted from Geller or others to the objects in question. With this realization came the parallel one that there were now hundreds of subjects claiming Geller's powers. Suddenly I appreciated that the time was ripe, at least for me, to make a serious study of these people. Many had come forward, so there would be no dearth of subjects, one of the problems of past investigations into the paranormal. Nor were the phenomena of spoon bending and watch starting easily lost in difficulties as to whether they really occurred or not, since one should have a bent spoon or ticking watch at the end of the test. Even more important, one should be able to observe the phenomena in action—in particular, the spoon in the process of bending.

It was that last feature that spurred me on. If I could obtain documentation of spoon bending, say on videotape, why could I not place various sorts of detectors between the subject and the spoon to observe the energy causing the spoon to be bent? Such a study began to take shape in my mind, with the realization that there was a lot of scientific work I could do on the problem. Science was not licked yet! More hopefully still, I might find an explanation as to how the phenomena occurred which would still fit into a suitably extended science.

The decision to undertake such an investigation was not taken lightly. As I remarked earlier, there are the two extremes of fanatical believers and outright sceptics, who attack each other virulently. In between these two warring camps, each bristling with its own pet armor and hurling abuse at the other, stands the defenseless but interested observer. He himself has a particularly thankless task to perform to discover the truth. Certainly he must avoid at all cost irrevocably joining either of the two fortresses. However, since he has to be able to discover in detail what resides in each, he will have to enter them to converse with their inhabitants. He will find it difficult to pass freely from one camp to the other, and indeed if he ever enters one camp he may be barred from approaching the other. And, from my own experience, whatever steps he takes will be criticized by them both.

Besides expecting no thanks from either the believers or sceptics, I realized that I should not expect much help from my academic colleagues. Numbers of them had already expressed displeasure at my appearance on the "Talk-In" program with Geller and others soon expressed hostility toward my attempts to start an investigation into the phenomena. I also knew that very little financial support would be available from the usual fund-giving bodies. Nor was the necessary apparatus or laboratory space easily available. However, I did find some staunch allies in my own college.

I was able to borrow part of a room and certain essential items of equipment for my various tests, as well as call on invaluable advice from several scientific colleagues as how best to use the apparatus in various situations. A young physics technician, Ray Ibrahim, was particularly helpful in this respect, and was closely involved in a number of the trials I carried out. Dr. Anthony Smith of the King's College engineering faculty, also with an interest in ESP, gave time and encouragement on metallurgical aspects of spoon bending. As the tests developed, various other people also gave much help, especially a young Argentinian physicist, Eduardo Balanovski, who had just finished his Ph.D. on the theory of the solid state. He was closely involved in tests performed both at King's College and at the homes of subjects in various parts of England. With small sums of money provided by private concerns to keep Eduardo from starving completely, I realized I would be able to proceed with the search for ESP.

My resolve was made even firmer by the various subjects I met informally. I became acquainted with them in numerous ways. Telephone numbers or addresses were made available to me by media organizations whom subjects had contacted when they had discovered their "powers." Others of them contacted me directly as a result of various media programs in which I appeared after the "Talk-In" show.

The first part of my project was to informally observe the subjects in action in their own homes, so as to assess the way in which I could best document the phenomenon. I decided to restrict myself to spoon bending alone, since that seemed the easiest to validate as occurring without fraud.

Initial results with various subjects, including Geller, left me in no doubt that there was something worth investigating. As one example, I have myself held the bowl of a teaspoon as its stem was rubbed gently by Geller till it broke gently into two pieces within a minute. The spoon, as well as I can recall the case without an exact videotape record, had come that instant from my pocket and I see no way in which it could have been tampered with beforehand. Excess pressure could not have been applied since I would certainly have sensed it as I held the bowl of the spoon. No duplication by an already bent spoon could have occurred since I held the spoon throughout until it broke, and in any case it was a rather special one with certain marks I could recognize. Nor could enough temperature have been applied to melt it, since if it had I would have smelled burning flesh—my own! The use of chemicals was ruled out by the absence of the characteristic discoloration and cracking observed in such a situation. In all, it was a case very difficult to understand unless I had hallucinated or not recalled all that happened correctly.

My resolve was then firm: I would attempt to understand spoon bending, and as scientifically as possible. I later hoped to use that knowledge as the clue to an explanation of the whole of the paranormal. I have often been told that I should not have wasted any time at all on the investigations I am describing here. I can only say in reply that if I could obtain in this way a proper understanding of the paranormal, then it would be worthwhile. In any case I was compelled to perform the investigation to rid myself of the feeling I described earlier. My hoped-for future explanation of spoon bending and other related paranormal effects would return me to the security of my own scientific background.

Beside my initial tentative investigations of the various available subjects, I also had to puzzle out how science itself could meet the challenge of the paranormal.

2

The Inexplicable Paranormal

BEFORE I BEGAN MY ACTIVE INVESTIGATIONS INTO PSYCHIC phenomena I had already become acquainted with the vast range that such phenomena can cover. Particular examples had caught my eye and seemed both reasonably well validated and difficult to explain in terms of present scientific knowledge. It seems appropriate to present some of these cases at this point to amplify the two features of authenticity and scientific impossibility. At the same time, these instances could give clues as to possible ways in which science may have to be extended in order to accommodate them.

I will start with spontaneous human combustion (SHC), in which people literally burn themselves to death, because it has been least remarked on among psychic phenomena while being about the most startling. One instance was introduced by Charles Dickens into the plot of *Bleak House*. The discovery of what is left of a sinister character appropriately named "Krook" is described in Chapter XXXII: "There is a very little fire left in the grate, but there is a smouldering suffocating vapour in the room, and a dark greasy coating on the walls and ceiling. . . . Here is a small burnt patch of flooring; here is the tinder from a little bundle of burnt paper, but not so light as usual, seeming to be steeped in something; and here is—is it the cinder of a small charred and broken log of wood sprinkled with white ashes, or is it coal? O Horror, he IS here! and this . . . is all that represents him." What makes this all the more interesting is a comment in the author's preface: "The possibility of what is called Spontaneous Combustion has been denied since the death of Mr. Krook; and my good friend Mr. Lewes (quite mistaken, as he soon found, in supposing the thing to have been abandoned by all authorities) published some ingenious letters to me at the time when that event was chronicled, arguing that Spontaneous Combustion could not possibly be. . . . Before I wrote that description I took pains to investigate the subject. There are about thirty cases on record . . ." —which Dickens proceeds to enumerate.

One of the best investigated cases of SHC concerned a 170-pound woman, Mrs. Reeser, who was last seen on a Sunday evening in July 1951 resting in an armchair in her flat in Florida. Eleven hours later all that was found of her was a shrunken skull, one vertebra and a left foot wearing the charred remains of a slipper. A report on the case by an investigator noted: "The windows of Mrs. Reeser's apartment were open all night. Yet no one smelled or saw any smoke. The apartment itself, though unbearably hot, was almost completely unscathed. The rooms were coated with oily soot above a four-foot line only. The armchair, lamp, chairside table and small portion of the carpet directly beneath the armchair were destroyed. Nothing else was damaged. It is known to require a temperature of 3,000° to so completely consume a human body. Mrs. Reeser's skull was shrunk to the size of a baseball, yet in all other known cases of intense heat the skull expands—sometimes even exploding." An arson agent who investigated the case stated: "We do not know that it was a fire—we just don't know what could have caused it." Nothing in the room was capable of producing such a high temperature. The local police and FBI investigated the case but were completely baffled.

There are a number of cases where it is clear that the source of heat is coming from inside the sufferer's body. In another reported case a woman found her invalid sister sitting in a chair and literally covered in flames. The woman wrapped a blanket around her sister, put out the flames and carried the badly burned woman upstairs. She then rushed out to fetch help, but when she had returned with a doctor her sister had been reduced to ashes with the exception of her head and fingers. Apparently the sister had caught fire again after having been extinguished. The bedclothes were supposedly undamaged, though there was the coating of greasy soot over nearby objects which was reported in the previous case.

Less extreme instances of psychic "heating" have been reported. One, which was well documented, concerned employees in a lawyer's office in the town of Rosenheim, Germany. It was related to the presence of a nineteen-year-old employee, Annemarie. A Miss Sch. and another employee sitting across from her complained of sudden strong pressures on their right or left ear and a strong flush reaching down to the neck, diagnosed as hyperemia. Miss Sch. also had contractions in her arms and legs. These effects ceased when Annemarie was absent, as did other psychic occurrences which will be described in more detail later.

Psychic healing is a related area, where healers apparently achieve miraculous cures on others, again by an act of will. Let me now quote some typical well-documented cases. The first involves the healer, Bill Clark, from Florida, who is sixty-six years of age and usually goes into a trancelike state during his healing sessions. In one such session, observed by a professor of pathology, the healer went into his trance with deep and rapid breathing while his eyebrows twitched uncontrollably. As he passed his hands over the backs of several patients his arms began to crackle as though with high voltage electricity. Even more strange was the initial purple discoloration of the skin on the healer's arms, which then turned into blood oozing from his pores, blood that was later tested and found to agree with the healer's blood group. Apparently the healer could raise the blood pressure in his arms by increasing his pulse rate to such an extent that the blood vessels ruptured and bled through his skin. It was as if he were sweating blood, though more descriptively his arms were "weeping" blood.

Such "weeping" of blood is very similar to stigmata which have been noted in various cases in history. A recent case is interesting because the woman, Maria Giorgini—whose stigmata appear on her hands and feet on Good Friday—discovered that she could heal. Doctors who worked with her at her clinic in the small village of San Baronto in Italy have testified that she has cured cases of diabetes, sterility and even lung cancer. One Florentine neurologist announced that he "can personally testify that seventy-year-old Teodore Todaro of Florence was cured of lung cancer by Maria." Nearly three years ago X-rays at the Careggi Hospital in Florence showed that Todaro had a lung tumor the size of a small ball. Doctors said there was no hope for his survival without an operation, but he was afraid to undergo surgery. Instead he went to Maria every week, and after six months—when doctors said he should have been dead—new hospital X-rays revealed that the tumor was only the size of a pinhead. Today Todaro is much better.

In another instance of healing, the astrologer Sybil Leek had been crippled for months and in almost unbearable pain from an ulcerated leg caused by an infected insect bite. The leg had swollen very considerably and she required a wheelchair or crutches to move around freely. She attended a conference in Hawaii while in this state and met a medicine woman called Moana. According to Sybil Leek, the treatment from Moana was as follows: "She held my hands and then went into a period of deep meditation. Suddenly I felt the pain flowing out of my leg and I felt strength in my leg for the first time in months. There was a strange electrical force to Moana. I felt her energy whenever she touched me. On that day I threw my canes away. I walked barefoot on the beach and galloped for miles. It was a wonderful, remarkable experiment. We agreed to meditate separately at 10 P.M. no matter where we were for the next two weeks. My leg continued to improve dramatically until my leg ulcers healed—all without hospitalization and drugs. By the end of the second week I was completely cured." A doctor who saw Mrs. Leek both before and after her remarkable recovery said, "She was in very severe pain and had to use crutches at first and then canes. I don't honestly know what happened when she went to Hawaii, but she showed an incredible improvement on her return."

In yet another, a Frenchman named Serge Perrin was suffering badly from thrombosis and his doctors expected him to die shortly. Perrin even made his own funeral arrangements. In a last desperate attempt to save his life, Perrin's wife convinced her husband to make the pilgrimage to Lourdes. On May 1, 1970, during the ceremony there of anointing the extremely ill, Serge fell unconscious. Yet after he had been anointed by the priest Serge recovered dramatically. "Suddenly my ears opened. It was like a cork being pulled from a bottle," Perrin recalled. "I felt a tingling warmth in my left leg—which was strange since I'd had no sensation in my leg for months. Soon the same thing happened in the other leg." After the service Perrin was able to walk, and his sight was totally recovered by the next day. He soon recovered totally, in a way baffling to the French medical profession. The doctor who had been treating him is reported as having said of Perrin, "His case was hopeless. . . . We knew we could not help him. His recovery cannot be attributed to medical intervention."

Finally, let me remark on the American, Edgar Cayce, who died in 1945 and who gave many "readings," as his diagnoses were called.

His wife's aunt was diagnosed by the established doctors, who wanted to operate on her, as having a tumor of the abdomen. Cayce was called and in a reading gave the information that there was no tumor but that his relative was pregnant and also suffering intestinal difficulties. The doctors regarded this as very unlikely since the woman was not supposed to be able to have children. Yet her husband, also a doctor, agreed to try the prescription which involved various enemas and other remedies. These were applied and the woman rapidly felt better. She gave birth seven months later to a baby boy.

Cayce in his waking state said he had no knowledge of anatomy or medicine whatsoever. A hospital anatomist said of one of Cayce's early readings that it was a perfect diagnosis and that he knew the patient. The doctor added that Cayce's anatomical knowledge was flawless.

I shall now turn to the "faculty" of clairvoyance, that of seeing things happening—usually to loved ones—far off.

In 1863 a Mr. Wilmot of Bridgeport, Connecticut, wrote: "I sailed from Liverpool for New York on the steamer 'City of Limerick.' . . . On the evening of the second day out . . . a severe storm began which lasted for nine days. . . . Upon the night of the eighth day . . . for the first time I enjoyed refreshing sleep. Toward morning I dreamed that I saw my wife, whom I had left in the U.S.A., come to the door of the stateroom clad in her nightdress. At the door she seemed to discover that I was not the only occupant in the room, hesitated a little, then advanced to my side, stooped down and kissed me, and quietly withdrew.

"Upon waking I was surprised to see my fellow passenger . . . leaning upon his elbow and looking fixedly at me. 'You're a pretty fellow,' he said at length, 'to have a lady come and visit you this way.' I pressed him for an explanation . . . and he related what he had seen while wide awake, lying on his berth. It exactly corresponded with my dream. . . .

"The day after landing I went to Watertown, Connecticut, where my children and my wife were . . . visiting her parents. Almost her first question when we were alone was, 'Did you receive a visit from me a week ago Tuesday?' 'It would be possible,' I said. 'Tell me what makes you think so.' My wife then told me that on account of the severity of the weather . . . she had been extremely anxious about me. On the night mentioned above she had lain awake a long time thinking about me, and about four o'clock in the morning it seemed to her that she went out to seek me. . . . She came at length . . . to my stateroom. 'Tell me,' she said, 'do they ever have staterooms like the one I saw, where the upper berth extends further back than the under one? A man was in the upper berth looking right at me, and for a moment I was afraid to go in, but soon I went up to the side of your berth, bent down and kissed you and embraced you, and then went away.' The description given by my wife of the steamship was correct in all particulars, though she had never seen it."

An oft-cited clairvoyant dream is that told by the Reverend Henry Bushnell in 1875. It occurred to Captain Youatt, who in a dream saw a company of emigrants perishing in the mountain snow. He could even make out the faces of the sufferers and specially noticed a perpendicular white rock cliff as part of the scenery. He fell asleep again and the dream was repeated. He described the scene to a friend who recognized it as being the

Carson Valley Pass, 150 miles away. A company was collected with blankets, provisions and mules. On arriving at the Pass the emigrants were found in exactly the state portrayed in the dream.

Related to clairvoyance is telepathy, the "ability" of transmitting one's thoughts to others. An American woman named Mrs. K. E. Mershon lived on Bali from 1931 to 1941 and found there strong evidence for telepathy. She wrote recently about the Balinese: "ESP is part of their everyday life. If somebody wants you to come to their house, they'll 'call you on the wind' as the natives say. I first discovered this when I went to a village one day and found the people very irritated with me. 'Why didn't you come sooner?' they asked. I explained that nobody had told me to be there and they said, 'Oh yes, we sent for you.' I insisted that no one came to my house and they said, 'Oh, we didn't mean that.' They had sent for me mentally—with ESP—and were annoyed that I didn't get the message." Mrs. Mershon then said she had become proficient in using ESP, which she used to communicate with a fisherman called Yoman. "When tourists wanted to go out in the coral reefs I would just think, 'Yoman, come in, Yoman, come in,' real hard. His arrival would depend on how far out to sea he was when I 'called.' If he was close to shore he'd come in immediately." Others who had lived on Bali supported this. A Cornell University professor who had lived on Bali for many years added about the Balinese: "They believe these powers are partly a natural gift but that they can be partly accentuated by meditation and spiritual discipline."

Precognition, the power of foretelling the future, is something many of us would like to possess. An interesting case of precognition has been put forward by the American psychic researcher W. E. Cox. He wrote: "On October 2, 1968, I was working in my shop. The radio was running on WPTF-AM, Raleigh, North Carolina. . . . About 1 P.M. I heard a newscast of the area, and a statement about the death of a Baptist minister associated with the Baptist State Convention in Raleigh. A mail truck had collided with his car. I heard this news item repeated two hours later on WPTF, and I heard it a third time shortly after 6 P.M." After being away for a day, Cox telephoned WPTF the morning following to question them about the newscast, since he had been surprised it had been given three times with no modification. The radio presenter replied, "I didn't give that newscast myself; but it so happens I do remember, since I was here. We learned of it about 11:30 P.M. It was aired on our midnight newscast; then we repeated it later yesterday morning." Cox replied, "I had thought I heard it the preceding afternoon; in fact I feel sure I did, and more than once," but received the response, "No sir, I don't think that would have been possible. I'm sure that it wasn't, because it happened late that night—about 10:30 I think." The accident had happened at 9:50 in the evening of October 2, and accorded closely with what Cox had apparently heard earlier. Nor could he have heard any news of the accident during the day he was away, according to his own account.

A more powerful psychic ability is that of psychokinesis (PK) in which objects such as billiard balls, matches, etc., are moved around purely by an act of will of the subject. An American parapsychologist, Pamela Painter de Maigret, visited the Russian subject, Alla Vinogradova, and saw her in action while being filmed in a TV studio in Moscow in 1973. "Alla appeared to be under considerable tension despite her bright warmth. As the filming began she rubbed her hands together, then reached toward a round aluminum cigar tube which lay near one edge of the table. She stopped her hand about six inches from it. Staring at it she raised and lowered her hand quickly several times, never coming closer than five or six inches from the tube. Then with a look of immense strain and concentration she turned the palm of her hand forward and appeared to 'push' at the cylinder. It rocked back and forth several times, then slowly began to roll across the plastic tabletop away from her hand. When the cylinder neared the outer edge, Alla quickly reached over the tube and put her hand on the far side of it. It stopped abruptly six or eight inches from her hand, rocked a few times and then started to roll in the opposite direction. Alla's hand was always at least half a foot away from both the table and the cylinder."

The writer of that report found that she herself could achieve the effect though only for a limited time. After she had been helped by Vinogradova, the report continues, "Alla stepped back and I continued to roll the cylinder across the table. When it came to the edge I copied Alla and put my hand on the far side of it. It stopped, hesitated, and then started rolling away from my hand."

A poltergeist ("noisy ghost") is a very close relative of PK, wherein objects are moved, even thrown, around a room apparently without anyone causing them to or even so willing them. In 1967 strange disturbances began to occur in the offices of a lawyer with the name Adam in the Bavarian town of Rosenheim, as mentioned earlier. In his own words: "It started very early, at the beginning of the year 1967, with telephone disturbances, with short circuits in the wiring. . . . Later on the striplights turned out of their fittings by 90 degrees and they have partially fallen down, sometimes exploded on the ceiling. We asked the Electricity Board to come, we asked the Stern Co., a well-known firm— everything to no avail. Then the striplights were replaced by hanging lamps, big globes. Here as well the lamps have broken into pieces, have burst. . . . I was almost a wholesale customer for striplights and bulbs, things happened so often. Then fresh events happened. Later on the lamps were swinging, they were swinging by themselves, swinging up to the ceiling. Then later on, the pictures on the wall turned themselves; all the pictures here in this room, including this very expensive picture. It turned itself, the picture fell off the wall, the wire has broken off as if it was cut off, as if it got broken. Still later, drawers were coming out of the writing desks by themselves. This cupboard, a 350-pound oak cupboard moved itself eighteen inches away from the wall on one side. Two policemen were able to move the cupboard back into its old position only by exerting all their strength. In the afternoon of that day the cupboard moved a second time."

This wide range of poltergeist-like effects were first investigated by various electricians called in by the lawyer on his suspicion that there were faults in the power supply system and the telephone connections. This latter was a particularly perplexing question because of excessive telephone interference, frequent failures of lines obtained and many calls to the time clock. The telephone bill was far larger than Herr Adam had expected, yet the telephone company tested the installation in the lawyer's office and found no faults with it in any way. The problems continued, so that finally the usual telephone installation was removed and replaced by a single telephone with a counter to record which telephone numbers were dialed and when.

The list of calls made during October and November turned out to be absolutely astounding. Certain numbers were dialed again and again. On October 5, the number 08036 282 was dialed four times in the minute starting 11:32 A.M. as was 02821 3027 at 3:19 P.M. On October 9, 30313 was dialed four times at 11:04. But the most astonishing feature was telephone calls to the time clock, this being the number 0119. On October 17, that number was dialed the staggering total of eighty-five times.

3

The Framework of Science

THE PHENOMENA DESCRIBED IN THE PRECEDING CHAPTER HAVE all been authenticated to an extent that makes them worth considering seriously. All remain to some degree unexplained. Clearly much work needs to be carried out if such things are ever to be understood properly. Although no single explanation need be expected for the whole of the supernatural, we might find features in common for at least some of the phenomena.

For instance, there does seem to be a similarity between the cases of spontaneous human combustion, production of psychic burns and psychic healing. We might conjecture that all involve an as-yet-unknown source of energy in the human body which is stronger in some people than others. The energy is released in a controlled manner by a psychic healer at a beneficial level to his patient. Psychic burns are produced when the energy release is too strong, while spontaneous human combustion would occur if the energy source became completely uncontrollable, thereby destroying its possessor. The idea of an energy supply in the body might also be relevant to the cases of psychokinesis (PK) and poltergeists described in the previous chapter. For PK the source would be under its owner's control, while for poltergeist activity it would decidedly not be so. Nor might the person causing the poltergeist effects have any knowledge that they possess such an energy source.

The examples of clairvoyance and telepathy cited in Chapter 2 could also involve some sort of energy transmission and reception peculiar to human beings. But the energy source itself would have to be sensitive to the thoughts of its possessor in order for such complicated information to be transmitted to someone else. The recipient of a telepathic message or a clairvoyant would also need to have high sensitivity to this unknown information-bearing energy emitted by others.

Precognition would seem to come under a different category from the other cases considered so far. Information is being received by a "sensitive," so we could suppose that it is being carried as packets or waves of energy in some manner. But this energy would have to move around in ways presently unknown to science.

Let us consider the above ideas in a little more detail. As we have suggested, it only seems possible to explain the terrible cases of spontaneous human combustion as being caused by some internal source of heat over which the individual has no control. Moreover, this source must be deep inside the body and would most likely reside in the bones, in order to reduce them to ash. It is difficult to conceive of any way in which this could occur unless there were storage of radiation in the skeleton as the energy built up to a dangerous level. The bones would act as so-called "dielectric resonators," the radiation bouncing back and forth inside them until it got so strong that it consumed them. It is still difficult to see in this theory how the bones could be reduced to ash, though if their temperature were high enough initially they might not crumble until it were too late for the energy just to escape by leaking out. Be that as it may, we have arrived at a conjectured mechanism for spontaneous human combustion, which is that people can, in as-yet-unknown exceptional circumstances, effectively become microwave ovens and burn themselves to death in the process.

Naturally enough, it is difficult to be in the vicinity of someone undergoing spontaneous human combustion, and it is very unlikely that even if one were—and equipped with suitable sensitive radio detectors—one would waste time "listening in" to any possible radio-wave emission instead of trying to put out the flames. In any case there is some evidence that indicates that SHC could have a simpler explanation. It may arise from highly combustible gas accumulating in the subject's body which accidentally becomes ignited.

There are two cases to support this view. One is of a clergyman who was unable to blow out the candles after a church service since his breath would burst into flames. He was successfully treated for this inflammable condition. Another led to a recent court case in Europe where a veterinarian was called in by a farmer to see his cow, which had a distended stomach. The vet put a hollow tube into the cow's anus, and tested the gas being released by applying a light to it. According to the records, the gas caught fire and acted "as a flame thrower," thereby destroying a considerable portion of the farm. If there is a possibility of such combustible products being formed internally, it is feasible that they could build up to such proportions as to cause self-destruction.

But are we making a mistake at this point in trying to give a scientific type of explanation of paranormal phenomena? I have already remarked that if they exist, then psychic events contradict science, so in order to understand them we should surely concentrate on developing an alternative to science. This idea of "alternate science," as it is called, has had many sympathizers recently, particularly among the younger generation who often regard science as having failed them by causing pollution and desecration of the environment as well as creating the atomic bomb.

It is essential to remember one point: science has progressed considerably since the first "noises" of psychic phenomena became of public interest in the middle of the last century. Through repeatable experiments a remarkable picture of the material world has been laboriously constructed by many patient scientists. This understanding should be used to its maximum in analysis of the physical features of psychic phenomena. While these latter may not be explicable simply in terms of modern science, and the majority of psychic events even appear to contradict present science, at least certain features of the phenomena involve matter whose behavior is expected to be governed by the laws of science. Thus a piece of crockery flying across a room during a poltergeist manifestation still appears to carry the energy it was imparted with as it thumps or smashes against the wall. A spoon bent "psychically" still appears to have gone through a sequence of plastic deformations before reaching its bent shape.

Even if the laws of science are being broken during paranormal events, it would be most valuable to find exactly which laws are being so violated and to what extent. This led me to the conviction that the apparatus of modern science should be incorporated as much as possible in the experimental phase of the search for the truth of psychic phenomena.

Scientific method itself has come in for much criticism recently. Science proceeds by isolating certain phenomena which can be made to occur again and again. Such repeatability allows the events to be investigated in an objective fashion, so that all who look at the phenomena carefully enough can ultimately agree as to what they see. A consensus can thus be reached about the phenomena. Explanations can be suggested and tested for their validity by going back to the events to ensure that the explanations really do fit the facts. For example, the trajectory of a projectile can be studied repeatedly to discover how the object falls toward the earth. Repeated analyses of such cases led to the discovery of the force of gravity acting on all objects near the earth, and imparting to them a downward acceleration of 32 feet per second per second.

The evidence presented earlier would appear to contradict modern science to such an extent that it might even be impossible to utilize it in the manner suggested above. If telepathic communication can be transmitted through lead shielding, it could not be explained by radio waves, the only explanation available to modern science (as we shall see shortly). Spoon bending achieved without contact seems hard to reconcile with the principle of conservation of energy, since a large amount of energy needs to be transmitted from the "agent" to the bent or broken spoon; a similar problem arises with poltergeist activity and willed object moving. Precognition is in complete contradiction to modern science, in which causality ("cause before effect") reigns supreme. Life after death is very difficult to conceive in an incorporeal fashion, while any continuation of a physical form after death appears extremely unlikely.

There is presently great sympathy for not attempting to analyze psychic phenomena. For example, the famous British faith healer, Harry Edwards, before he died recently, tried to discourage me from any physical investigation of faith healing. Nothing would ever be found out about it in that way, he advised. But if a cure is brought about in a psychic healing session, then some physical process must have occurred in the patient's body. It would be of immense value to humanity to discover what it was that had triggered off such an effect. On these grounds we should at least try to see how far the scientific method can take us in our search into the nature of parapsychology. We have already argued along such lines at the beginning of the book; if ESP does contradict science we should try to find out the exact nature of this disagreement.

In order to see the general form such a contradiction could take, let us turn to the question of how science works. A precise or scientific explanation of an event must be objective in the sense that there is agreement on the meanings of the terms being used. Otherwise scientists and others could not agree about or even properly understand what was contained in such an explanation.

So the first step in the process of explanation is that of description, and indeed this is an essential step. Without it no discussion can proceed satisfactorily. Then comes the step of relating the various parts of the description to each other and to other phenomena altogether—thus of seeing correlations. This is followed by the explanation proper, where various of the described properties are deduced from supposed underlying properties.

The process of explanation therefore depends on the description, since it is the described features which have to be explained. Description itself, to be objective, must be quantitative. Numbers, either integer, rational or irrational, must be associated with the various features for this to be achieved.

For example, let us consider the size of an object. It does not help to say the object is large or medium-sized unless one is given a standard of comparison. "Large compared to the size of my body" is a quantitative description at a very primitive level, since it involves only two or three categories or classes: objects either larger than, equal to or smaller than my body. More precision may be needed, such as that the object is between $2\frac{1}{2}$ and $2\frac{3}{4}$ times the size of my body.

Aspects of an object that would appear to be more subjective, such as its color or its texture, must also be described quantitatively in order for unambiguous description to be given. To say an object is colored blue might be very misleading if the person giving the description were color-blind. This could be avoided by saying that the object looked blue to most people, but there might be a high proportion of color-blind people in the population considered. The only objective way to specify the color is to give the range of wavelength it emits when illuminated in white light.

Quantification precedes scientific explanation, then, for any phenomenon. The next point to realize is that mental experience has never been quantified in an objective fashion. It is possible for me to claim that the toothache I experienced last night was greater than that I suffered the previous night. But there is no way in which this statement can be verified by others in the way that statements about sizes, etc., of objects can be. Indeed I may be wrong in my comparison of last night's and the previous night's toothache. I might have forgotten the excruciating pain I underwent on the night before last either for conscious or unconscious reasons. It would not seem possible for others to invade my private world of experience to check such statements or any others that I might make about my mental experiences.

If such invasion of the privacy of a person's mental experience were not possible, then science could be applicable only to material phenomena. Even more extreme would be the result that objective discussion about mental experiences would be impossible. Thus materialism—the idea that all is constructed from matter—would win by default, so to speak, since it would not be useful to discuss anything other than events associated with matter, as far as explanation were concerned.

This is an attitude which appears to be logically possible, though extremely limiting. It would not allow a scientific discussion of all phenomena, if we include as phenomena our own subjective mental experience. We must therefore attempt to extend materialism to our own inner experience. Our inner privacy must be penetrable by probing scientific tools if it is ever to be analyzed scientifically. There are already examples of this occurring in the use of, among other things, lie detectors and electroencephalographs, which assign definite numbers to mental states. The ultimate aim of such an approach would be to determine enough quantities associated with the physical activity in a person's brain to allow a complete specification of the internal experience. This would include the level of intensity of such experience.

Only by such an approach would the world of inner experience, or consciousness, emotion, sensation, etc., be amenable to a scientific investigation. In other words, *only in the materialistic approach is the mind accessible to science.* Any other model of all phenomena can only be limited to a scientific analysis of the material world. The rest is described in terms which could mean different things to different men. Nothing could ever be proved. Indeed, that is borne out by the philosophical meanderings that have been produced ever since the beginning of man's historical records. The mind/body problem can only be overcome by a materialistic approach.

Materialism does not deny consciousness, but tries to explain it in physical terms. That is not an easy thing to do, but there does not appear to be any reason in principle why it cannot be done. In any case it is necessary to attempt it; the parody of the materialist denying the existence of his own consciousness is too gross an affront to the facts of life to be acceptable. Sadly there has been little progress in setting up a materialistic model of consciousness. Yet slow but sure progress is being made in brain research in understanding the physical basis of the mind. A materialist approach need not be forsaken because of the difficulty of the problem of consciousness.

Certain features of the world may always be beyond our grasp. But then it is not clear that such features can even be known to exist. It could be supposed that some knowledge may be beyond verbalization. Against such a view, which is only really the extreme version of solipsism, we can only be pragmatic. In order to continue discussion we have to assume that we can discuss meaningfully with others. Otherwise we should be silent. Therefore we should assume the materialistic model as a possible one, and see how well it works. If it breaks down at some points, then we will be forced to retreat into silence. We should, however, see how far we can advance before our lips are sealed.

It is here that we come back to our earlier question about the nature of the possible contradiction between science and parapsychology. The materialist explanation of all phenomena would necessarily fail if certain of the latter were discovered to be unquantifiable, as is claimed by some in respect of mental events. Yet material effects must be associated with paranormal events, for otherwise they would not be noticed. Even in the case of an hallucination, definite brain processes are expected to occur as concomitants of such an experience. It therefore seems difficult to understand how at least a certain measure of quantifiability cannot be attached to paranormal phenomena. They may display certain vagaries, such as unrepeatability, that would cause their deeper scientific investigation and possible explanation to have great difficulties. But at least a start can be made on their quantification. The true nature of further problems hindering us in understanding parapsychology will only become evident once such a start has been made.

Before so doing we should realize a little about the detailed nature of the materialist explanation of phenomena given so far. The reductionist program, in which the properties of objects have been explained in terms of their constituents, and they in terms of their constituents, has been most successful. This chain has led to molecules, atoms, electrons and nuclei, and finally to the interior of the nucleus, which has been found to contain subnuclear particles with a bewildering number of companions.

Besides these constituents are the forces holding them together. Four are presently known, these being gravity, radioactivity, electromagnetism and the nuclear force holding the nuclear constituents together.

From this array of constituents and forces binding them to-gether only one force and set of constituents turn out to be a possible means of explaining the paranormal: electromagnetic forces acting on the normal constituents of our bodies. Electro-magnetism is the force which is the basis of chemical reactions in our bodies. Through purely human effort it is only such a force, acting on such constituents as our bodies are composed of, that will be available to us. At least millions of times more energy is required than is normally available to utilize one of the other three forces of nature.

It might be claimed that there is an unknown "fifth force" of nature causing paranormal phenomena and observable only when these phenomena are occurring. Such a force would have to transmit a great deal of energy, for example, to bend a spoon in the way achieved by Uri Geller. This energy would have to overcome the electromagnetic forces binding the metal atoms together. If the fifth force were other than electromagnetism the atoms would need to respond more strongly to the fifth force, while it is operative, than to electric forces. They should there-fore produce this fifth force themselves, as does an electrically charged particle produce its own electric field, a magnetically charged one its magnetic field, and so on. Such an additional force between atoms should therefore exist all the time, and not only during paranormal occurrences. There is absolutely no scientific trace of such a force in atomic physics, down to many orders of magnitude, in comparison with the electromagnetic force. We thus discard the idea of a fifth force if we wish to preserve the scientific viewpoint.

We therefore have to accept that when science faces up to the supernatural, it is a case of "electromagnetism or bust." Thus we have to look in detail at the various paranormal phenomena to see if electromagnetism can be used to explain them.

To do this we must search for an electromagnetic signal of suitable strength emitted by a psychic during the time he or she is causing a particular paranormal effect. Thus the validation program must involve electromagnetic or radio-wave detectors of the right sort—sensitive enough and with the right frequency range. If no signal can be found, then we shall have to take a very hard look indeed at the particular phenomenon. If we find no electromagnetic signal whatsoever—over the whole frequency range, and the complete gamut of supernatural phenomena— then we may be in a difficult position. Science and the supernatural will have come into a head-on collision.

It is exactly this search for the electromagnetic signal occurring during paranormal phenomena that I have made during the last four years. It is the results of this search that we shall have to consider shortly.

Radio waves can have a whole range of wavelengths—from kilometers down to centimeters, the latter being called "microwaves" because of the relative smallness of their wavelength. More energy can be transmitted by electromagnetic radiation as heat or infrared (as from the sun), where the wavelength tends to be much shorter (down to thousandths of a centimeter). We should remark that light is also a form of electromagnetic radiation with a wavelength of a few hundred thousandths of a centimeter. All of us emit electromagnetic radiation over the whole range of wavelengths (called "black-body" radiation), though most of the energy comes out as infrared. This can easily be detected by holding a sensitive thermometer near to (but not on) the skin. The thermometer will heat up as a result of the infrared rays falling on it from the nearby skin.

Human microwave emission can be measured by modern techniques, termed microwave radiometry. This allows the minuscule amounts of energy we each emit, at wavelengths of a centimeter or so, to be monitored. Since radio waves with wavelengths longer than about three centimeters can penetrate deeply into the body, it is possible to use this emission to monitor for "hot spots" inside the body. This might correspond to tumors, so the technique is clearly of importance. It has been used in the U.S.A. to look for early signs of breast cancer in a random sample of two thousand women; not a single one of these had microwave emission beyond a few percent above normal. What is expected, on our "human microwave oven" hypothesis, is many times more emission than normal.

Human radio-wave emission does not seem able to explain the poltergeist phenomenon since the normal amount of relevant human radiation is more than a billion times less than needed for such an explanation. It has not been possible to measure if there were excess microwave emission during SHC, but on the same grounds it is very unlikely to be the cause. In the same way, the "psychic burns" are unlikely to be explained in terms of excessive microwave emission.

It is also unlikely that such a mechanism is available to produce psychic raps which occur during table-moving sessions. Raps and clicks have indeed been produced in various pieces of plastic by directing microwave beams into them; people can even hear buzzes and hisses in their heads when standing in a microwave beam. I have not attended a séance where such raps occurred, but here again the level of microwave or other form of radio wave required to produce this effect would have to be a billion times greater than that produced normally by the human body.

As with poltergeists or table rapping, it would not appear to be possible to explain "table moving" electromagnetically. To effect the displacements, electrostatic fields would have to be so strong that it would truly be a "shocking" experience; radio or microwave beams would be expected to produce so much energy as to roast the assembled company.

Yet earlier we realized we were in the position of "electromagnetism or bust" with respect to psychic phenomena. We can only move forward on the supposition that the way radio waves interact with or are produced by the human body is stranger than anyone had ever suspected. We would only find these strange effects in special circumstances, in particular associated with those who claim psychic powers. We can but search for a concomitant radio signal along with the psychic results.

4

Psychic Healing

PSYCHIC HEALING DOES OCCUR. PEOPLE ALSO CURE THEM-
selves, and indeed hundreds of cases are known where spontaneous remission of cancer has occurred without outside intervention. Is that the answer to psychic healing? Does the healer,
purely by psychological means, bring out in the patient the will
to live or the wish to be cured, and the patient do the rest?

Faith healing in the West has important documentation in the
Bible. Thus: "And when he was come unto the house, the blind
man came to him: and Jesus said unto them, Believe ye that I am
able to do this? They said unto him, Yea, lord. Then touched he
their eyes, saying, According to your faith be it unto you. And
their eyes were opened" (Matthew 9:28–30). There are numerous cures of similar form, brought about by Jesus; among others
he is supposed to have cleansed a leper, cured a woman of a fever
and even brought back from the dead a young girl. Indeed Jesus
appears to have been one of the greatest, if not the greatest, of
faith healers: "And great multitudes came unto him, having with
them those that were lame, blind, dumb, maimed, and many
others, and cast them down at Jesus' feet; and he healed them"
(Matthew 15:30).

There are also those healers who not only require that religion
be involved but claim that guidance is given to them by the spirits

of the dead. Thus, Arigo, a Brazilian "psychic surgeon" who had been investigated by Western doctors before his death in 1971, could make remarkably accurate diagnoses of patients. He claimed that he achieved this by "listening to a voice in my right ear, and I repeat whatever it says. It is always right." This voice was supposed to be that of a deceased German medical student, Dr. Fritz. Such "spirit healing" performed with the help of spirit guides also occurs in the West with a very enthusiastic following.

Finally, there are healers who have no claim to communication with the spirit world nor profess great faith in religion yet who believe they cause healing by some special energy they radiate. They sometimes claim they can channel "cosmic" energy from outside themselves into the patient. Alternatively they may suppose they emit it from their own bodies. The descriptions of this emission vary enormously. One healer claims that his hands become "as heavy as concrete or lead" when he is being effective as a healer. Others sense a white light around themselves or feel intolerable heat in their fingers or hands. Some healers may need to touch the patient's body, though not necessarily the diseased portion. Other healers may try to affect the patient from afar.

How can we assess the validity of the different forms of psychic healing? This is a question which presents severe difficulties. Thus a patient may spontaneously recover from a disease which might be on the verge of killing him. It seems hard to differentiate such cases of spontaneous remission from cures brought about by the activity of someone else. Furthermore, the healer may help to heal the patient precisely because he gives psychological support to the latter and so rallies him to resist his disease. The psychological factor is as hard to assess in healing as it is in many other walks of life. The effect of the environment on the patient must also be considered, in spite of the lack of any reliable indicator to describe it.

Another feature essential in the investigation of psychic healing is the lack of detailed records in so many cases of treatment. A healer might claim that he has cured 30 to 40 percent of his patients of their ailments, as did the late Harry Edwards, the English faith healer. Yet such a claim is hard to establish. Detailed records of patients are few, especially with any degree of medical substantiation. Hostility to healers has always been shown by the medical profession. Supporting medical evidence both as to the nature of the disease being suffered and its subsequent progress is therefore scarce.

Many healers, especially in the Western world, also advise their patients to see or continue to visit their doctor. Therefore even if such documentation were obtainable, the help gained from medicine or other treatment prescribed by the doctor would thus have to be discounted in the process of cure before the healer's contribution can be assessed. It is possible to take the view that if healing occurs, then that is all that is needed and further investigation is unnecessary. This position, even taken by some psychic healers themselves, is clearly very shortsighted, since the discovery of the critical factors promoting healing would allow them to be augmented and healing could be accelerated.

Let us turn, then, to the process of psychic healing itself. At one extreme is so-called psychic surgery, in which the healer's hands appear to penetrate the patient's body without any knife being used. Diseased tissue is then removed by the healer, and as his hands are removed from the patient's body the hole closes up, leaving no trace of the opening. At the other end of the spectrum is distant healing, where patient and healer both sit at the same time to meditate on the patient's illness, though at some distance apart. Healthy thoughts are supposed to be entertained by both, the healer "willing" the patient better during the process.

Psychic surgery is clearly a very bizarre process indeed. It is practically impossible for a patient's body just to open up as the healer's fingers approach the surface of the skin. Neither can tissue be removed from the patient's body without surgery unless great damage is done, nor can holes in the patient's body surface close up, leaving no scar. Yet many "operations" of this type have been performed and even witnessed by Western observers; some are supposed to have been very successful.

An account given by a qualified medical doctor of such an "operation," performed by a Brazilian psychic surgeon, Zeca, is as follows: "I looked down at the patient, a man of thirty-eight who had been diagnosed as suffering from a heart condition, and I saw that an incision about fifteen centimeters long had been made right down his costal edge." Then Zeca just separated the flesh and plunged both hands inside, or rather just his fingers, and began to manipulate violently. Then he took a pair of pincers and pulled out a piece of tissue, which he dropped into a metal container. Next, he simply "pressed the two sides of the opening together—that's the only way I can describe it—and the skin just closed up." A similar, but now firsthand, account of such an operation on his stomach by the Brazilian healer Edivaldo was published by Guy Lyon Playfair: "Edivaldo's hands seemed to find what they were looking for, the thumbs pressed down hard, and I felt a distinct plop as they penetrated the skin and went inside. My stomach immediately felt wet all over, as if I were bleeding to death. I could feel a sort of tickling inside, but no pain at all."

The last account is most interesting, especially the fact that wetness could be sensed by Playfair on the surface of his stomach, as well as "tickling inside." If pain sensors had been inactivated by the healer in some way it would have been impossible for wetness or tickling to have been felt at all. For pain and other sensory receptors on the skin's surface have similar features, and it would appear difficult to have one active without the other. One possible way in which pain could be reduced is by direct action on higher control centers in the brain. This may happen when a patient is hypnotized and told he will feel no pain when his tooth is being drilled. A similar mechanism may be at work when acupuncture is used in surgical operations, though the acupuncture needles stuck into appropriate parts of the body are also playing an important role in keeping down pain, in addition to the strong suggestions made beforehand that the patient will feel little pain. However, from responses of hypnotized people, there appears to be a similar reduction in other sensations in the particular area of the patient's body surface, so that such an explanation does not appear feasible.

A simpler way of explaining the apparent opening up of the body in these and similar cases is that no hole is actually created in the body but only an illusion of it. The appearance of a cavity is particularly easy to simulate in the surface of the stomach since it is soft. If the healer's knuckles are bent over at the same time it would appear quite convincing that his hands have indeed penetrated the patient's body. The blood which usually accompanies such "operations" could have been concealed on the healer's hands. The "plop" sensed by Playfair could have been caused by a hidden container being burst by pressure on the surface of his stomach and so releasing a liquid looking realistically like human blood (even being animal blood, easily obtainable in agriculturally-oriented communities).

This explanation of fraud appears highly likely in a good proportion of such "psychic operations." One Western observer was even told by a leading psychic surgeon that at least 99 percent of such "operations" were fraudulent but appropriate for the rather gullible populace. Other observers have had their suspicions aroused when they have investigated closely, though no clear evidence of cheating has been observed. But at no time have such "operations" been performed under sufficiently controlled conditions for them to be taken seriously.

We can only conclude that there is no real case for psychic surgery. There are many less "miraculous" operations reported, such as those with Arigo, the so-called "surgeon of the rusty knife." Another firsthand account of a typical operation, which was also being filmed, was published by the medically trained psychic researcher Andrija Puharich: "Arigo with a flourish requested that someone furnish him with a pocket knife and someone in the audience produced one. Arigo took hold of my right wrist with his left hand and wielded the borrowed pocket knife with his right hand. I was told not to watch the operation on my arm, so I turned my head toward my cameraman and directed the motion picture work. The next thing I knew was that Arigo had placed a tumor and the knife in my hand. In spite of being perfectly conscious, I had not felt any pain. In fact, I had no sensation at all at the surgical site. I was sure that I had not been hypnotized. Yet there was an incision in my arm, which was bleeding, and there was the tumor from my arm. Subsequent analysis of the film showed that the entire operation lasted five seconds. Arigo had made two strokes with the knife on the skin. The skin had split wide open, and the tumor was clearly visible. Arigo then squeezed the tumor as one might squeeze a boil and the tumor popped out." The incision did not close up immediately, but the wound healed in three days, about half the time it would normally have been expected to take. Nor was there any infection in spite of the pocket knife and Arigo's hands being dirty during the brief operation.

However, these features of lack of infection and rapid healing are not universal, and at least a third of such "dirty-knife" type of operations are known to become septic. Psychological effects may also be present to speed up wound healing.

Western psychic healing involves either the "laying on of hands" or "distant healing." In the former practice the patient sits or lies in a relaxed state, sometimes in trance. The healer gently passes his or her hands near or over the patient or lays them on the patient's body, especially the diseased portion. Distant healing has the patient and healer meditating for a few minutes at the same time each day. There are numerous healers who intersperse the occasional "laying on of hands" treatment with distant healing. There are also cases where distant healing alone is attempted.

Let us first consider a typical case of faith healing brought about by a woman healer whom I have investigated in the south of England. The patient was a middle-aged woman who had suffered very badly from arthritis for the previous eighteen years. The pain in her knees and ankles had become so intense that she could no longer climb the stairs of her house or hold a saucepan. She had been required to take about fifteen aspirin a day to reduce the pain. She was chosen by a drug company to appear in a film indicating how a victim of severe arthritis could still cope by using certain methods. On her first visit to the healer she was not sceptical but open-minded. The healer ran her fingers down the patient's spine, causing "hot spots of pain" to be produced. At the end of that first session all pain had been removed from the patient's back. Marked improvement was noticed within three weeks, during which further treatment occurred; soon thereafter the patient stopped taking her aspirin and certain other drugs, and within a year was able to go dancing and function normally as a housewife in her own house.

We cannot deduce with certainty that it was the healer alone who caused the clear cure in the above case. The patient was still on certain drugs for a time and these might have been of help. However, established medicine had not achieved any improvement of the patient's condition until the healer was consulted. It seems reasonable, therefore, to conclude that the healer played an important role in the case. However, we cannot assess the true role of the healer for the reasons given earlier: we do not know how much psychological help was given, as distinct from some definite physical energy.

In a series of tests on psychic healing which I carried out with E. Balanovski, the relationship of the patient to the healer turned out to be highly variable. Three healers were used to treat one particular patient, each healer working with him for an hour, with a short rest before the next healer commenced. The relation of the patient to the healers was extremely negative in one case, neutral in another and very positive in the third.

The second feature which supports the theory of a psychological basis of faith healing is the insistence by the healer of a short period of meditation by the patient at the same time each day. The healer usually encourages this by saying that he or she will be sending out healing energy at that time. This meditation period was noted in Chapter 2 in the case of the healer Moana and patient Sybil Leek. Even without energy transfer such meditation is undoubtedly beneficial.

There is a relation between such a meditation period and a third piece of evidence which comes from another area of fringe medicine gradually being incorporated into the establishment, that of biofeedback. This involves monitoring various physiological parameters of the patient, such as skin resistance or temperature, and requiring the patient to learn how to modify these to his or her advantage. Indeed the range of illnesses for which such techniques are effective has proved quite large, including control of epilepsy, removal of migraine headaches, reduction of stress-producing muscular pains and the lowering of blood pressure.

A detailed case of blood pressure control was reported at the 1974 conference of the British Society for Psychosomatic Research by Dr. Chandra Patel. She had carried out various studies of patients which showed that biofeedback techniques were effective. In one study, twenty patients being treated were matched to twenty hypertensive controls for age and sex. At the end of the treatment the average blood pressure of the treated group was reduced from 160/100 to 140/85. The average blood pressure of the control group was reduced from 163/99 to 162/97. In addition, there was an average reduction in antihypertensive drug requirements of 41 percent in the treated group.

Biofeedback has also been used with apparent success in curing crippling back pains. A California hospital has used such techniques, together with psychological counseling, and helped a number of otherwise intractable cases. A bus driver who had damaged a vertebral disk in a crash, and had been treated unsuccessfully by traditional methods for the resulting unbearable pain, was given relaxation biofeedback training and psychological counseling. The patient learned how to relax his body and increasingly reduce his pain.

Another feature apparently of great importance in the healing process is the will to live or die. This often determines whether a seriously ill patient lives or dies. People with a strong will to live have been known to survive fatal illnesses for years. A good example was the divorced woman with leukemia who insisted she had to live until her son was old enough to care for himself. To everyone's surprise she lived thirteen years longer and did not die until soon after her son decided to get married. She lived for thirteen years beyond what was expected; her commitment to her son kept her alive. The essential feature of this type of healing was put succinctly by the scientific director of the Psychosomatic Clinics at the Maryland University School of Medicine: "You cannot overestimate the biological power of one human being on another." The relevance of this to faith healing is clear.

Also important is the so-called "placebo effect," which involves giving innocuous medicine (which is claimed to be highly effective) to patients. They are so impressed that they can be cured. Up to 40 percent of the population will get relief from pain when given chemically innocuous tablets which are supposed painkillers; the person most helped being the extrovert who wants to help his doctor help him more. Conditions as serious as heart disease can be improved by placebos, and even organic diseases such as cancer might be helped by them. A patient with a great will to live may be able to use this placebo effect and become cured, even of a terminal illness, because of his great belief in the healer's powers.

These pieces of evidence give considerable support to the thesis that psychic healing is achieved by purely psychological means in the mind of the patient. Yet there are claims of "electrical" energy flowing from the healer to the patient, as was reported in Chapter 2. Sensations of heat or cold are experienced by the healer and patient during certain healing sessions. Furthermore, recent techniques have been developed that are claimed to be able to measure the "psychic energy" of a healer during the healing process; these go under the name of Kirlian photography. Ways of measuring the "healing energy emission" have also been claimed. It is of value to discover if there is indeed such an emission. This I have tried to do.

Perhaps there is more to psychic healing than a purely psychological explanation. This would involve energy of some sort being transmitted from the healer to the patient during a healing session. Attempts to detect this energy have been tried in various ways, according to the form the energy might be expected to take. Such energy might, for example, take the form of an electromagnetic wave emitted by the healer in some as-yet-unknown manner. This wave could be either a very long radio wave, a much shorter radio or television wave, or one with a wavelength of only centimeters—the microwave radiation mentioned earlier. There are other parts of the electromagnetic spectrum which have been investigated scientifically, such as infrared, visible light, ultraviolet, X-rays or gamma rays. The first three of these can hardly penetrate the skin, so it would be difficult for a healer to emit them from inside his body. The other two would in any case be impossible for a person to emit utilizing only ordinary chemical energies.

Intense beams of radio, television or microwaves can be detected easily enough with suitable apparatus. However, it is not clear what sensitivity the human body is expected to have to such waves. Diathermy uses waves of ten-centimeter wavelength to cause heating in deep tissue. It may be that there are nonthermal effects of such or similar radio waves, so that it is necessary to be prepared for all eventualities. To do that it is necessary to measure the minute amounts of radio waves that each and every one of us emits because of our normal temperature (discussed in the previous chapter). This "black-body" radiation, as it is called, can be measured easily enough by its heat effects when it takes the form of infrared or heat waves. Thermographic cameras have been constructed which can take a picture of the infrared emission from a body. This is particularly useful in detecting tumors, by observing the local heating due to excessive cellular metabolism produced in the region near the tumor.

The detection of very long radio waves from the body has also proved possible, particularly because bodily processes produce far more radiation of this sort than the extremely low level of "black-body" radiation expected. The action of the heart in pumping blood around the body produces electric field variations of fractions of a volt which can be measured by detectors placed close to or on the body surface. There are concomitant magnetic field changes which can also be detected even at a distance from the body, though they may be only a millionth of the earth's natural magnetic field. There are also electrical changes associated with brain activity which are monitored by pasting electrodes to the scalp and amplifying the resultant signals. The concomitant magnetic fields have also been measured, in spite of being less than a hundredth of that caused by the heart.

There is currently a great deal of interest in discovering if microwaves or radio waves can be effective in the healing process. Various scientific groups have presented evidence that nonthermal effects do arise, such as the regrowth of cut nervous tissue, when low-level microwaves are directed at living systems. There is also a great deal of evidence from Iron Curtain countries that there is a higher level of cardiovascular and psychological disorders among workers in the microwave industry than among the general public. However, none of this is yet thoroughly substantiated.

Attempts are currently being made to see if small quantities of static electricity on various parts of the body are related to health or disease in related organs. Careful measurements were made in the earlier part of this century by a former anatomy professor from Yale University, H. S. Burr, which claimed to show that electrical potential differences between various points on the surface of the body could variously establish the precise amount of ovulation in a woman, detect malignant tumors and ascertain many other features of the state of health or otherwise. This work has since been continued in much more detail by Dr. Motoyama of the Institute for Religious Psychology in Tokyo. He claims to have related various types of disorder with the imbalance of electrical potential between various "seiketsu points" on the fingertips or toes where acupuncture lines begin and end. He has even claimed to alleviate suffering by restoring the correct electrical balance in certain patients by suitable electrical means.

We can conclude that a healer could achieve cures by emitting suitable microwaves and so help the patient in an as-yet-unknown way. Alternatively, he might alleviate suffering by modifying any imbalance of electrical potential between various parts of the body, for which there is some slight evidence, though again an absence of any detailed mechanism. Either of these processes might be used to explain the complete alleviation of pain by Arigo and its partial reduction by other healers. It might also give an understanding of cures such as the one of arthritis mentioned earlier. Such a task would be very difficult, but it is not impossible in principle.

In collaboration with E. Balanovski I tried to discover if there are any abnormally high electric or magnetic fields of very long wavelength near a healer's body during the course of a healing session. We also attempted to observe any abnormal radio-wave or microwave emission, working with a number of healers both from England and America.

Measurements of natural microwave radiation proved hard to make because the apparatus needs to be extremely sensitive to measure the low levels. Groups in Canada, America and Sweden have carried out measurements on human emission over a wide variety of wavelengths, but apparently only when the subjects have been in a normal state. We used similar techniques; at lower frequency we were able to utilize readily accessible equipment to perform the detection.

Our measurements were taken as the healer was attempting to heal a patient. Both healer and patient were invited to the laboratory and first spent a little time there to accustom themselves to their surroundings. Various healers and patients were investigated in such a situation, sometimes the healer and patient having worked together extensively before, in other cases being strangers to each other.

The healer and patient would then be wired up with various electrodes (usually on the healer's hands) if they were needed for the electromagnetic detection tests. The healing session would then commence and would usually proceed for about half an hour or more. Various movable detector aerials might be brought up to the healer's or patient's body during that time and recordings taken of electromagnetic emission at various wavelengths (to be compared with those taken before and after the healing session).

At no time was an electromagnetic signal observed beyond the normal levels to be expected. At the very low frequency end of the spectrum we did not attempt to test Dr. Motoyama's theories because it was impossible for us to beg or borrow the very expensive apparatus required. But we did note that there were no dramatic changes of body electric potential of either the healer or the patient. Thus we concluded that electromagnetism was not involved in healing.

This left somewhat of a puzzle, since we had not been able to explain the sensations of heat or cold reported as experienced in many healing sessions. We decided that we should make sure that heating or cooling really did occur in the faith-healing situation. We therefore measured the surface skin temperature of the healer and patient during a test, using electric thermometers at the same time as their subjective temperature sensations were noted. The results showed quite conclusively that the experienced changes of temperature were purely in the mind and had no basis in fact. The healer or the patient might suddenly exclaim, "I feel my hands are getting hotter—they're hotter—they're very hot," and then a few seconds later say, "They've got cooler." Yet there was no such change of temperature, even to a tenth of a degree centigrade. There was thus no paradox, since there was no actual tissue heating in either healer or patient.

There have been other attempts to detect healing energy, supposing it does exist. Healers have claimed to modify the structure of various substances, especially that of water. This has then been given to patients or poured over seeds to enhance healing or growth. It has even been claimed that such "treated water" has different physical properties, such as viscosity, surface tension and other physical parameters, compared to the untreated variety. My colleague E. Balanovski and I have investigated water treated by a variety of healers and have detected no difference between it and distilled water. We have only found that experiments repeated often enough to give good statistics indicate no effect. Experiments of this type which have been claimed to detect "healing energy" require careful repetition before acceptance, with all suitable variables controlled. The same must be said about the claims that high voltage or so-called Kirlian photography can be used to detect the healing energy. This involves photographing the sparks emitted from an object when it is placed in a strong enough electric field.

Kirlian photography was first investigated by the Russian, Semyon Kirlian, in 1939, and its use has since spread widely throughout the world. Great claims have been made on its behalf in detecting "psychic" energy in a variety of situations. Thus Dr. Thelma Moss has claimed that "we were able to corroborate what the Soviet literature had reported, that different states of emotional or physiological arousal in the human being would reveal very different patternings in these photographs." It even seemed as though parts of a leaf, for example, would still be visible by Kirlian photography even though they had been torn away. These results were used to support the idea of the existence of a "bioplasma body" or "etheric body" around a living organism. That inanimate objects could be observed by this technique led to the idea of stones, etc., having their own "etheric bodies."

Careful work has now been done to test the various claims put forward by the believers in the efficacy of Kirlian photography. Two different groups in America, one at Drexel University, the other at Stanford University, as well as my own group at King's

College, including Eduardo Balanovski and Ray Ibrahim, have attempted to perform Kirlian photography under as well-controlled conditions as possible.

There are various features of the process which require care and attention. One is the amount of pressure applied by the subject to the photographic plate or other surface. It is well known that under differing psychological states a person can apply varying amounts of pressure to a surface, so that must be controlled. A second quantity with which care must be exercised is the angle which the fingertip from which the sparks are being drawn makes with the surface on which it rests. In the extreme case, even portions of the human fingernail may be captured on the photograph if the finger is tilted up too much. The moisture on the fingertip is also of importance, as is the length of time the finger is exposed.

When all of these factors are carefully controlled, there is no change of the Kirlian photograph with the psychological state of the subject. It seems that the most important variable is the moisture content of the fingertip, and the Drexel group even concluded that "corona discharge (Kirlian) photography may be useful in detection and quantification of moisture in animate and inanimate specimens through the orderly modulation of the image due to various levels of moisture."

Nor is there any question whatsoever of this technique actually measuring any psychic energy. A careful calculation by American scientists has shown that the process occurs by a well-understood physical process called a corona discharge. We can only conclude that Kirlian photography is irrelevant to paranormal investigations.

In spite of these results the gullible will still believe the impossible and improbable. They will still follow the newspaper reporter who wrote recently, "Every human being has an aura which [one] can actually see with Aura goggles (£4.60) or photograph with a Kirlian camera (£72 to order)"!

Finally, we come back to answer the question we asked somewhat earlier: there is no trace of any energy of any sort emitted during the healing process. This does not mean that healers are fraudulent; all the ones I have worked with were very conscientious and dedicated people. Further investigation should be carried out into the phenomenon if only to understand in what way the healer interacts with the psychological makeup of the patient to help him to heal himself. As far as the "bedside manner" is concerned, psychic healing has no doubt a lot to teach established medicine. But it is *not* paranormal.

5

Clairvoyance

A CASE OF WHAT SEEMS LIKE PURE CLAIRVOYANCE CAME TO MY knowledge in 1974 when I visited Andrija Puharich at his home in Ossining, New York. He and his colleagues had been putting "sensitives" into hypnotic trances and asking them to give answers to questions which science presently had been unable to solve. One of these sensitives, a young woman, had been asked in a trance state to describe the internal structure of the neutron, the other constituent of the atomic nucleus beside the proton. This is a problem which scientists are very concerned to solve since it should give an understanding of the hundreds of other particles, short-lived companions of the neutron, which have been discovered since 1945.

The young woman's description of the inside of the neutron was quite dramatic. She told me what she remembered of her trance experience. She recounted how she had floated through a haze to approach a large silver globe which had a porthole in one side. As she floated in through this opening she experienced a slightly nauseous sensation in the pit of her stomach. Inside the globe she saw the sudden appearance of star tracks all departing from a given point. They would vanish only to have similar lines of stars diverging from another point suddenly appear elsewhere in the globe. After observing this for a short time she floated back out of the silver globe, again experiencing the nausea in her stomach as she went through the porthole. Nearby was an almost identical silvery globe, again with a porthole. However, she had no sensation as she floated through that opening, nor did she see the lines of stars inside. That, she indicated, was an empty neutron.

It must have been pure coincidence that ten days previously I had visited the National Accelerator Laboratory at Batavia, Illinois, where is sited one of the biggest and most expensive atom smashers in the world. There were various experiments being set up or performed which I was interested in visiting. One of these was to discover the internal structure of the neutron. A beam of very energetic neutrinos, particles of no electric charge or mass, was being used. Neutrinos can be considered as pure energy, so they can act as probes inside the proton and neutron. In this particular experiment the neutrino beam was shot at a large silver globe with a porthole in its side, through which the neutrinos entered the globe. The globe itself is called a "bubble chamber" since it is filled with a suitably unreactive liquid such as propane. It is operated at a pressure such that it is just about ready to boil. If a proton or neutron is split apart by a direct hit from a neutrino, the fragments flying off cause a chain of gas bubbles to be formed. Light is shone into the liquid from its side so that the bubbles look like a line of stars. There would very likely be a number of these lines diverging from the point of fragmentation of the original neutron or proton. By the side of the filled bubble chamber being used in the experiment was another, almost identical to the first, but completely empty.

When I heard the young woman's account of the interior of the neutron I was astonished at the almost complete similarity it possessed to the experimental arrangements for investigating the neutron's inside. In any case, her description of the interior of an actual neutron was unlikely to be correct. Yet what would be more natural than that she would be receptive to experimental attempts to probe inside the neutron?

The only question to be raised against this case as one of bona fide clairvoyance is the possibility that the young woman had been able to discover the experimental setup herself. She might have read a popular article on the subject or talked to someone about it. It was possible that she could have read a *Scientific American* article or something similar, describing how atom smashers work, and how subatomic particle fragmentation can be observed by photographing tracks in bubble chambers. Not that she possessed any conscious scientific knowledge of this sort; but the subconscious may be able to store a great deal which could be accessible when in a trance state, such as achieved by hypnosis (evidence of this is presented in Chapter 10). Even so, she could not easily have learned about the use of neutrino beams at the National Accelerator Laboratory.

These cases did not occur under the sort of laboratory conditions in which most scientists feel properly at ease. This lack of control means that no case can finally be accepted as perfect evidence. If clairvoyance truly does occur, then it should ultimately be possible to capture it in the laboratory. The claim to have done so already has been made by a number of investigators in the past, particularly the American parapsychologist J. B. Rhine and his colleagues at Duke University, in Durham, North Carolina. They performed extended series of tests during the 1930s and 1940s, with some remarkably successful outcomes.

Rhine and his colleagues attempted to answer the following question (as formulated in their important book *Extra-Sensory Perception After Sixty Years*): "Is it possible repeatedly to obtain results that are statistically significant when subjects are tested for knowledge of (or reaction to) external stimuli (unknown and uninferable to the subject) under conditions that safely exclude the recognized sensory process?" Their investigations covered both clairvoyance and telepathy, but we will delay considering the latter faculty until the next chapter.

The clairvoyance tests were conducted with the use of a specially designed pack of ESP cards. The pack consisted of twenty-five cards divided into five equal sets with one each of five different pictures—a circle, a square, a cross, three wavy lines and a star. The investigator would shuffle the pack and place it face-down. The subject would then try to guess the order of the cards in the pack. He or she would do this either one card at a time as they were drawn successively from the top or bottom of the pack by the experimenter, or would indicate the order of the whole pack. The guesses or "calls" would be recorded, as would the order of the target deck.

Early tests involved rather few safeguards against conscious or subconscious fraud, and were criticized accordingly. By 1940 the major possibilities of error had supposedly been avoided. Mistakes in the statistical analysis of the data were claimed to have been effectively removed, and the American Institute of Statistical Mathematics issued a statement in 1937 to that effect. Inaccuracies in the recording of results were reduced to an insignificant level by having more than one investigator check the matching of scores and also by preventing the subject and the investigator from knowing the scores which had been obtained in any one test as it was occurring. Several copies of the original records were also made to prevent them being tampered with. Picking up cues from the investigator, for example, or by marks made on the cards, was also avoided by the subject and the cards being separated by a screen or kept to different rooms, if not different buildings. Shuffling defects were also shown to be absent by careful cross-checks on the order of the cards being guessed.

Several tests of clairvoyance were reported which had none of these various defects but yet had results at a significant level. One of these was the test carried out by J. G. Pratt of Rhine's group on the subject H. Pearce. Pratt and his subject were in different buildings one hundred yards apart. At agreed times Pratt displaced the cards one by one from the shuffled deck and Pearce independently recorded his guesses. At the end of the pack Pratt recorded the order of the cards. Copies of the records were immediately sent in sealed envelopes to Rhine, before the lists were compared. The total number of guesses was 1,850, of which one-fifth, or 370, would be expected to be correct by pure chance. However, the actual number of successes was far above that, being 558. This could have occurred by chance less than once in a hundred million similar trials.

These results were later criticized by C. E. M. Hansel, who pointed out that Pearce could have gone to Pratt's building and looked through the fanlight of Pratt's door to see the cards Pratt was handling or the score sheet. That Pearce could have attempted this is true, but further analysis claimed to have shown that neither cards nor score sheet would have been visible from that position. In particular the American parapsychologist Ian Stevenson claims that Hansel based his conclusion on an inaccurate diagram of Pratt's office.

Even more outstanding success was claimed by Dorothy Martin and Frances Stribic at the University of Colorado from 1938 to 1940. Various subjects were tested and one in particular, C. J., produced an average score of 6.89 per pack during a set of 25,000 guesses, so successful as to be completely insignificant by mere chance. The subject did not handle the cards, these being on the other side of a screen from him. One test with C. J. involved ten packs of cards, but he was only asked to guess the order of the cards in one of them. After guessing in this way for 110 packs, it was found he had scored an average of 8.17 per pack. Yet if his guesses were applied to the other nine packs his average score was only 5.02 per pack, as close as one would expect to the pure guesswork value of 5.

Tests of a similarly high level of success were performed by other workers, in particular by Pratt with a subject, Mrs. M., who could have achieved the results by chance only once in ten million times, and by Pratt and Woodruff with sixty-six unselected subjects and a total of 96,700 guesses. In the latter test, the results achieved could be expected by chance less than once in a hundred thousand million times. A further test involved only 250 guesses, with the subject locked in a separate room on a different floor from that of the investigators. The subject scored 9.3 successes for each pack of twenty-five cards, the odds against this happening by chance being astronomically high.

It would seem that any one of these various results (and others not mentioned here though discussed in the books referred to in the Further Reading section) constitute grounds for authenticating clairvoyance. They even appear to justify the claim that nearly everyone has clairvoyant powers. Such a result might also seem to be supported by the supposedly high incidence of clairvoyant and telepathic powers among primitive peoples, but this latter has not been tested under similarly rigorous conditions.

However, even if the charge of fraud in any one of these many successful tests were disproved, it is still hard to accept the data, based as they are on the laborious piling up of tens or even hundreds of thousands of guesses to magnify a supposedly real but very small effect. There is, in particular, one feature of the results which feeds the lingering doubt. It is the so-called "decline" effect, where a subject's score in card guessing may initially be high but decreases steadily to chance level over a series of tests. A description of this for one subject is revealing. "Near the end of the series the subject's performance declined. She was then tested on a less rigorous procedure in the hope of helping her initial level of performance, but this was to no avail. She was then tested again for 13,700 trials. . . . This produced only chance results. Nevertheless, if the results of these two blocks of testing are pooled, the results are still highly significant."

Even though a subject's initial success might be very high indeed, does the decline effect indicate that he or she had "beginner's luck"? In other words, is the initial success purely by chance, so that if the trials were continued long enough the early success would be swamped by later events? Such a criticism could be made to stick only if there had been inaccurate use of statistical assessment as to how likely a particular set of successful guesses was by pure chance. This had already been discussed by Rhine under the criticism: "The ESP (clairvoyance and telepathy) results represent simply a rare instance of luck; for example, like the one-suit bridge hand." As Rhine notes, I think correctly, "This hypothesis with the various versions of it are easily recognized as 'common sense' denials of the efficacy of the mathematics of probability."

Given that clairvoyance could be occurring, there are a whole list of questions which require investigation, some of which are:

a. What can be sensed clairvoyantly—pictures, or words as well, and what range of discrimination occurs? Can musical notes be sensed? Colors?

b. What is the dependence of the faculty on distance between viewer and viewed?

c. What happens if various shields are interspersed between subject and object?

d. Is there a critical dependence of clairvoyance on the size of the viewed object? For example, is there a critical size of pictures on one of the standard ESP cards below which clairvoyance disappears?

e. How fast can information be obtained by the faculty?

f. What part of the perceiver's body can be shielded to destroy the clairvoyant success?

The answers to these questions would allow an evaluation of the feasibility or otherwise of clairvoyance in terms of established science. Thus clairvoyance might be found to occur even if the cards being guessed were, for example, always in a metal box screening out *all* forms of radio waves (except very energetic X-rays). Under such circumstances, clairvoyance could not be explicable in terms of electromagnetism, a very interesting result if true.

There have been tests that have apparently indicated other features of the faculty of clairvoyance; one of the most valuable was made by H. E. Puthoff and R. Targ at the Stanford Research Institute (SRI) in California. They asked various subjects to perform "distant viewing" of targets in the San Francisco Bay area. The subject was closeted with one of the investigators, and waited for thirty minutes before beginning a description of what he considered was the chosen remote location. This latter was picked randomly by someone otherwise not involved with the tests from a pool of one hundred such locations known only to that person and within thirty-minutes driving time from SRI. Other investigators (two to four of them) involved in the test then drove to the appointed place and remained there for fifteen minutes after the allotted thirty-minute driving time. During this period, neither the subject nor the investigator with him had any knowledge of the target location; the subject described his impressions of the place into a tape recorder and made whatever drawings he thought appropriate.

There were six subjects involved altogether, three who had

experience of attempts at distant viewing and three who were considered as learners. One of these was Pat Price, a former California police commissioner. He seemed to be able to obtain remote information in his distant-viewing tests. In one case he recognized the target location and named it. Yet he had errors in some of his accounts. In a typical case he correctly described a parklike area containing two pools of water, though he thought they were used for water filtration rather than their correct use as swimming pools. He also included in his drawing some tanks which were not present at the target site.

The drawings and descriptions for a given subject were judged by an SRI employee not otherwise involved with the test. He visited each of the nine target locations and ordered the subject's various responses on a numerical scale as to degree of agreement (without knowing which response was to which place). In Price's case this ordering established that Price's descriptions and the actual targets they were aimed at were in remarkable agreement. It could have occurred by chance only once in thirty thousand similar cases. One of the learners was even better, with results obtainable by chance only once in half a million similar cases. She was adjudged to have made five direct hits and the remaining four guesses were very near misses.

Another feature relevant to the question of feasibility was that apparently there was only a slow rate of transmission of information. Thus: "Curiously, objects in motion at the remote site were rarely mentioned in the [subject's] transcript. For example, trains crossing the railroad trestle target were not described, though the remote experimenter stood very close to them." If clairvoyance was being used here, it could not allow any rapidly changing phenomena to be transmitted.

We are now armed, at least tentatively, with various features of the powers of clairvoyance. From the Rhine tests we know that quite small objects, down to $\frac{1}{16}''$ can be observed (even down to $\frac{1}{100}''$ in some cases). According to the SRI tests only a limited rate of information transfer can occur. Some of the SRI tests were conducted with the subject inside a metal chamber shielding out radio waves, unless they had very low or very high frequency. The results were as good as if no shielding had been present. This eliminates such a range of radio waves as carriers of clairvoyant information.

If clairvoyant viewing does occur, we would expect it to do so by means of some sort of wave being used by the subject to give a picture of the scene. Either sound or radio waves would be feasible. They could be used in an active fashion, in which case the subject would emit the waves and detect how they are scattered back to him from the objects which he is attempting to detect (as does a bat with its echo-location). On the other hand he might operate in a passive mode. In that case the subject would just sit back and pick up the waves coming to him (as happens with our ears and eyes). Sound waves seem to be ruled out over the several kilometers distance which occurred in various of the tests, for buildings do not emit them (except at very low intensity with very low energy), nor do people if they are in the active radar mode. This leaves only the possibility of radio waves being used. Again, since buildings do not emit any radio waves, only the active radar mode of distant viewing seems possible.

If such high accuracy can be obtained that objects only $\frac{1}{16}''$ in size can be observed, then the wavelength of the radio waves cannot be much larger, otherwise the waves would bend around the object to be detected, and so miss it. This requires that the corresponding frequency of the radio waves be high, about ten billion cycles per second (higher than television transmissions at about six hundred million per second). This frequency is difficult to reconcile with the apparently very low rate of signal transmission found in the SRI tests and also with the shielding (unless there was a bad leak in the shielded room).

Even if these discrepancies are discounted, there is a further problem. Human beings are not television transmitters emitting easily measurable amounts of electromagnetic power at such a high frequency. There is indeed natural emission of this sort from any object because of its natural temperature—the "black-body" radiation described earlier—though this can be detected only with very sensitive microwave detectors, mentioned in Chapter 4. Can this so-called "black-body" radiation be used so that a person acts as an active radar set? For this to occur the human body must be sensitive to very low power levels of such radio waves, even far below those emitted by other people nearby.

The initial reaction to this suggestion is that it is very unlikely that the human body could be as sensitive as is required to explain clairvoyance. We would all have noticed this long ago, and indeed would be expected to suffer dramatically as we approached a television transmitting station. It would appear that clairvoyance cannot be explained on the high frequency radio-wave model. Yet is it still possible that a passive radar mode is used where the investigator's emission is picked up by the subject from the distant target site, and then other nearby emissions are added to this? This is more feasible than the earlier model. However, there will still be at least a millionfold reduction of the signal by the time it is picked up several kilometers away. Therefore it is unlikely that any radio-wave model is viable for clairvoyance, though it is not completely ruled out on this argument.

There is a further aspect of the radio-wave mechanism for distant viewing which we have to think about with care. As I have said, radio waves can bend around objects of a size much smaller than their wavelength. Because of this they make such small objects appear "fuzzy." It therefore transpires that the further away an object is, the larger it must be for it not to be so blurred as to be unnoticeable.

We can apply this criterion to the results of the distant-viewing tests at Stanford or of the card-guessing tests reported earlier in this chapter. We must also impose the criterion that the wavelength of the radiation being used is at least 1 centimeter, since otherwise it would be involved solely with the body surface. We can then calculate that objects 4 kilometers away will be noticeable only if they are at least 60 meters high; at a distance of 100 meters they have to be at least 1½ meters in size. Each of these figures contradicts the reported results in the SRI viewing tests by a factor of about 1,000, in the card-guessing tests by a factor of about 100.

There have been further attempts at duplicating the SRI distant-viewing tests with apparent success. One of these involved participants communicating through a computer network. The participants were as far apart as Florida, California, New York and Quebec. They attempted remote viewing of various mineral samples. Their results were claimed to exclude chance successfully by odds of 25 : 1. Because of the distance of the participants from the target and the size of the target, the success contradicts the radio-wave hypothesis by a factor of at least 40,000.

We have to add to this series of disagreements the fact that in the Puthoff and Targ tests a double-walled, copper-screen Faraday cage was used to give electrical shielding from low frequency radio waves in certain of their trials. Such a cage would also shield out microwave radiation. Thus we can safely assume that electromagnetism has nothing to do with distant viewing.

Other forces of nature and their associated rays—radioactivity, the nuclear force or gravity—could also be considered as candidates for clairvoyant "rays," but can be rather quickly rejected since they cannot be controlled by the body. At this point, therefore, we are faced more forcibly with the problem raised earlier about clairvoyance: a miracle or lies? We will now retreat from a further confrontation with the impossible until we have looked at the parallel phenomenon to clairvoyance, that of telepathy. We might get some indication there which would help us break the deadlock in which we have found ourselves.

6

Telepathy

ONE OF THE FOUNDERS OF THE BRITISH SOCIETY FOR PSYCHICAL
Research, Frederick W. H. Myers, coined the word *telepathy* to
describe the extrasensory perception of another person's
thoughts. The word itself is derived from the Greek roots *tele* or
"distant" and *pathe* meaning "feeling" or "occurrence." There
are other terms describing this same phenomenon, such as the
word *biocommunication,* used by the Russians.

The Russian neurophysiologist, Professor L. L. Vasiliev, Professor of Physiology at the Institute of Brain Research, University
of Leningrad, experienced a very strong telepathic experience in
his own family when he was a child. He himself nearly drowned
when twelve years old. He had climbed along a sloping willow
tree which extended over a stream. He slipped and fell in, and
being a nonswimmer just managed to save his life by catching
hold to the tip of a branch. He lost a much admired new white
school cap in the process. His mother was ill at that time, being
with his father at Carlsbad, a long distance away. Vasiliev's young
aunts, in whose charge he had been placed, agreed not to inform
his parents. They were dismayed when the parents returned from
Carlsbad and told them about the boy's near drowning, including
the details about the willow and the loss of the school cap. Apparently Vasiliev's mother had awoken after she had a dream about

the incident and had demanded that her husband send a telegram to make sure that all was well. The telegram was never sent, though Vasiliev's father had left the room and pretended to so do.

The transference of thoughts from one person to another has been claimed to occur over very long distances indeed, even as far as around the earth. There is a report by the American astronaut Ed Mitchell claiming some proficiency in sending telepathic messages from the moon to certain subjects on earth. If these various results are indeed correct, then telepathy must be distinct from the possibly subconscious observation of nonverbal cues. The existence of telepathy between people who do not know each other well would also discount the tendency of people who have become familiar with each other to think alike. In other words, telepathy would on these accounts appear to involve perception other than by one of the known five senses. It would be paranormal.

Telepathy and clairvoyance are usually bracketed together under the term ESP. This combination is a very natural one since, in particular, it is difficult to distinguish the one from the other. Any telepathic success by one person of another's thoughts could as well be ascribed to the percipient's clairvoyant perception of the physical brain activity of the other person. This may still be possible even though scientifically we cannot discern a person's thoughts from the activity of his brain cells, except in a very general fashion by such indicators as EEG activity. Indeed, we perform many acts of will every day of our lives without having any understanding at all of how such commands are carried out by our neurophysiological machinery.

On the other hand it is difficult to distinguish a clairvoyant perception of a distant location from the telepathic probing of someone else's thoughts, and especially memories, of that place. This possibility naturally arises in assessment of the SRI results on distant viewing presented in the previous chapter. The place being viewed was always one associated with a number of people in the target team. Thus human involvement was essential, yet there was an indication that certain information was obtained by the distant-viewing subject which could not have been known to the target team. However, the perceptions being obtained could have been in the memory of one member of the target team. We cannot, therefore, separate clairvoyance from telepathy with any surety.

It is certainly harder to conceive of a simple mechanism which will allow telepathic perception of memories of a distant site to be obtained from people who are not at that place at the time that clairvoyance tests are being conducted. Yet we cannot exclude such a possibility since we are presently not clever enough to understand how such a phenomenon might occur. This problem is different from that of breaking the light barrier (that no one can travel faster than light). The experimental verification of the principle of special relativity indicates exactly the way the light barrier could be broached (if at all), in particular by a very large expenditure of energy. We have presently very few such quantitative limitations in the case of telepathy. Not that we cannot develop them, but we need to learn about the nature of telepathy before we can do so.

It is usually claimed that telepathic powers are prevalent among primitive peoples, and have only been lost as civilization developed and the abilities atrophied. There are communities—such as the island of Bali, mentioned in Chapter 2—that apparently possessed such powers until very recently. We might expect that the greater the emotional involvement of the transmitter and receiver, the greater the success. There are various cases of apparent spontaneous telepathy brought about by an accident happening to a friend or relative. An example of this was reported by a Russian sailor: "While serving on a submarine, I became ill and the ship had to leave without me. During an afternoon nap I had the following dream: I was right back on the submarine, standing on the deck. The boat began to descend into the water, but I was unable to reach the conning tower and make my way down into the safety of the ship. I was overwhelmed by the water, began to swallow it and felt that I was drowning. At this point I awoke, sweating and with my pulse racing. I remembered the dream quite vividly afterward. When the submarine returned to its base and I rejoined the crew, I heard that one of my comrades had drowned. He had accidentally remained on deck while the boat submerged. When I checked the ship's log, I discovered that the accident had happened at the very moment I experienced the nightmare of my own drowning."

Several studies have been made of spontaneous telepathic cases. One of them is by J. B. Rhine's wife, L. E. Rhine, who found that the majority of them were from women and that most of the events described were disasters or crises of one form or another. A careful analysis was made by the American psychologist Ian Stevenson of thirty-five telepathic cases. He found six of these were related with death and twenty with illness or accident; only the remaining nine cases had happy or neutral content. The information could be received while the percipient was awake or in an altered state of consciousness, though it was most impressive in the latter state.

These spontaneous telepathic cases are of value in indicating the most suitable environment in which to investigate more controlled transmission of information in this manner. They do indeed suggest that emotionally charged material be attempted to be transmitted, and some tests have been tried using such a technique. Before we turn to describe that, we should consider first the earlier work on card guessing especially associated with J. B. Rhine and his colleagues, for as in the clairvoyant cases the amount of material and its success rate is quite staggering.

We have already discussed the remarkably successful ESP experiments carried out by J. G. Pratt with his subjects Pearce and Mrs. M., the Martin-Stribic tests with the subject C. J. and the Pratt-Woodruff tests. All of these can be clearly identified as tests of clairvoyance as distinct from telepathy since no one, neither investigators nor subjects, were supposed to have had any conscious knowledge of the order of the cards in the pack until the order had been guessed by the subject. Other tests were performed in which one of the investigators did look at the cards while the subject was guessing them. Any success in these cases could be achieved by telepathy as well as by clairvoyance, as we discussed earlier in this chapter.

One of the most remarkable cases of card guessing ever recorded was reported by the American parapsychologist Dr. B. F. Riess. His subject, a young woman music teacher, wrote down her guesses at her own home over a quarter of a mile away from Riess as he turned over the cards in his home. There were 1,850 guesses (seventy-four runs of the customary twenty-five-card ESP pack), with the remarkably high average score of 18.24 per pack. This is the largest average score ever obtained by a subject. In a sequence of nine packs there were more than twenty successful guesses in each of them. The odds against this being by chance are ten, followed by seven hundred zeros, to one. This case was carefully analyzed by Rhine and his collaborators and shown not to be vulnerable to the various criticisms of nonrandom shuffling of the cards, sensory leakage in the experimental rooms, incompetence of the experimenters, errors of recording the targets and guesses, dubious statistical methods to analyze the data and so on. A total of thirty-five criticisms in all were considered, as they were for the clairvoyant card-guessing tests described in the previous chapter. As in those cases, Riess's tests apparently satisfied each of these attacks. Yet after criticism of the fact that the records were left by Riess in an unlocked drawer of a desk accessible to friends of the subject, Riess admitted: "In view of the uncontrollable factors, the data as presented are to be thought of as suggestive only." As Hansel concludes: "Thus, this experiment cannot now be considered as in any way inexplicable."

As in those cases, we can only conclude that either a miracle has occurred, or that fraud has been perpetrated by the investigators or subjects. Further analysis as to the feasibility of the phenomenon may help to clarify the situation a little, but before we can do that we need to learn more about other features of telepathy. In particular, tentative answers to the questions asked about clairvoyance in the previous chapter would be of great value here: How does telepathy (if it exists) depend on the content of the message being sent; on the distance between percipients; on shielding between percipients, especially of various portions of their body; and is there a limit on the rate of information transfer by this means? To give some indications of the answers to these questions, we will have to turn to more recent work, where conditions have been more varied.

To begin, however, it is possibly of value to remember a case of spontaneous telepathy reported in the *Proceedings of the British*

Society for Psychical Research and verified by a number of independent witnesses. The report is as follows: "On September 9, 1848, at the siege of Mooltan, Maj. Gen. R., then adjutant of his regiment, was severely wounded, and thought himself to be dying, and requested that his ring be taken off and sent to his wife. At the same time she was in Ferozepore (150 miles distant), lying on her bed between sleeping and waking, and distinctly saw her husband being carried off the field, and heard his voice saying, 'Take this ring off my finger and send it to my wife.'" This case is of value since it indicates that a number of sensory modes may be excited simultaneously by telepathic signals, even over hundreds of miles. It also indicates that a reasonably high rate of information transformation would be needed in order for the auditory information actually to be audible; the auditory information appears to have been received at about the normal rate of speaking.

We will turn now to the more recent work where attempts have been made to send emotionally laden material to subjects. Some suggestive results have been obtained by the American parapsychologist, Thelma Moss, when senders viewed slides and listened to tapes with strong emotional content. The response of one receiver to a slide-tape presentation concerning the assassination of President John F. Kennedy was as follows: "I seem to have the feeling of sadness or sorrow . . . as if I were crying . . . or something tragic has happened, and that I was grieving over something . . . much the same as one might feel attending a funeral of a dear friend . . . or a well-known figure in whom one had faith. . . ." A group of twelve judges matched the responses of thirty senders and thirty receivers in these tests. Out of these, seven were able to achieve results significantly greater than chance. However, the level of significance was in no way comparable to that achieved in the card-guessing tests described earlier. Still, Moss's tests give support to the idea that emotionally charged messages containing a considerable amount of information can be sent telepathically.

If telepathy and/or clairvoyance—lumped together as generalized ESP, or GESP—were a faculty possessed by man in his past which has now been lost by evolutionary developments, it might still be possible to catch a glimpse of it in early childhood. This suggestion has been put forward seriously by Ernest Spinetti, who conjectures that "the ability to employ GESP declines as the human being learns more and more to employ and generate individually specific cognitive relations." Spinetti has tested this idea with schoolchildren of ages from three to seventeen years and with older people as well. The subjects were divided into age groups with a spread of several years in each group, except for the two youngest groups who had age ranges from 3.2 to 3.6 and 4.5 to 4.9 years, respectively. A hundred subjects were chosen from each age range and divided into pairs. One of each pair was asked to send a picture to the other in an adjoining booth. After ten such trials their roles were reversed for a further ten trials.

The results were exactly as Spinetti had expected. The three youngest groups, in order of increasing age, had a success rate of 451, 361 and 261 in one test, where 200 was expected by pure chance. The remaining groups had no significant difference from 200. In another series of tests with a slightly better setup, the corresponding numbers were 226, 174 and 122, the remaining older group scoring very close to the pure chance value of 100. These results could only have happened by chance less than once in a thousand similar tests. We might note here that one of the reasons for the apparently satisfactory completion of the tests with the youngest children was the use of a "games-playing" approach to them.

Many attempts have been made to discover if there is any appreciable falloff of telepathic powers as the distance between the sender and receiver increases. The longest distances have been involved with the Apollo 14 moon flight, when E. D. Mitchell attempted to transmit pictures of ESP cards at prearranged times to four chosen subjects back on earth. Because of Mitchell's heavy duties, only half of the tests could be carried out, and the results of these were not highly significant. Other tests have been attempted by the American parapsychologist Karlis Osis and also by Russian scientists. All of these results seem consistent with the hypothesis that telepathic (and clairvoyant) powers decrease with distance (though are not conclusive enough to prove this case completely). This might be expected if there is a physical process involved in the phenomenon. We might consider a telepathic sender emitting bursts or pulses of energy coded, for example, by Morse. A receiver would be able to notice these signals provided they were above the minimum level to which he is sensitive. This is the so-called noise level, where all the other signals impinging on the receiver, constituting noise, swamp the signal. Such a model would allow for transmission of telepathy out to a certain distance and no further. Near the end of this range it would become harder for the signal to be picked out of the noise, but this cutoff might occur quite rapidly. We can only conclude that results on the falloff of telepathy with distance are certainly of interest, but cannot easily be predicted.

A series of tests which are supposed to contradict any radio-wave explanation of telepathy completely were directed by Professor L. L. Vasiliev, whose own experience with telepathy has already been mentioned. An Italian neurologist, F. Cazzamali, had earlier suggested that telepathy was achieved by radio waves. He claimed to have discovered such waves emitted by subjects attempting telepathy. These waves were supposed to have had a wavelength from about 1 to 100 meters and corresponding frequencies from three million to three hundred million cycles per second. Vasiliev's work in telepathy was mainly devoted to establishing the truth of Cazzamali's ideas.

Vasiliev's work was performed in the 1920s and 1930s. The most important result was that lead shielding around the receiver in no way interfered with his or her reception of a telepathic signal. This has often been regarded as a complete contradiction of Cazzamali's radio-wave ideas, since such waves would not be expected to penetrate the shielding used in the experiments. Yet Vasiliev wrote: "From this it must be concluded that, if the transmission of thought at a distance is effected by radiation of electromagnetic energy emanating from the central nervous system, then such electromagnetic energy must be sought either in the region of kilometer electromagnetic waves or else beyond the soft X-rays, but neither supposition is at all probable." Because of the importance of this result, let us look in detail at what the experiments comprised.

After a number of initial tests, the technique was developed of putting a subject to sleep or awakening him or her by a telepathic message. This was a method which had been used by others apparently with significant success. The technique Vasiliev used was to attempt to make the subject fall asleep and, if successful, to try to wake the subject again after a short while, each time by telepathy. The subject's state of alertness was measured by his ability to give a rhythmic contraction to his hand, which was automatically registered in an adjoining room. As the subject dozed off the signal would slow down and ultimately cease. On waking, the signal would recommence.

Three different subjects were tested as to their responsiveness to distant commands to sleep or wake. Several hundred tests were performed, some of these without the sender or receiver being screened, others with either sender or both sender and subject being in screened chambers. The chambers were shown independently to be impervious to radio waves from 1 to 100 meters in length generated by a small radio transmitter. The results of these tests showed that the length of time it took from the beginning of mental suggestion by the sender to the actual sleeping or waking of the subject was independent of the presence or absence of any metal screening. In another set of tests the average time from the beginning of the experiment to the onset of sleep was about seven minutes when no telepathy was being used and reduced to half that when mental suggestion was attempted either with or without metal shielding. This supposedly showed that mental suggestion was operative and not influenced by the metal shielding.

A number of criticisms can be made of the experimental method used in these various tests, though the majority of them were answered in Vasiliev's text published in 1962. I think we can accept as reasonable that there were no possibilities of simple auditory or other types of cues between sender and receiver. There were certainly successful tests in which there was a distance of several kilometers between them. There are two crucial features which must be examined, however. The first of these is the question as to how significant the results really were. There were considerable fluctuations of the time between the onset of telepathic transmission and the subject sleeping or waking. Thus, in the important case of tests on four subjects reported by Vasiliev, they were put to sleep by telepathy in about one-third of the time they would normally take in falling asleep on their own. But there were large variations in the latter, very nearly as large as the shortening in time to fall asleep brought about by telepathy. A similar feature holds for the other results, and we can only conclude that there is not a clearly significant effect.

The other question was already remarked on in the quote from Vasiliev's report. Very low frequency radio waves were claimed to be excluded, but no reason at all given for that claim. Indeed it has been proved possible to observe low frequency radiation at a distance from the body, and even from the head, of normal people. Thus, this remark of Vasiliev is not correct.

There is also the question as to the level of radio waves which might penetrate the shielded enclosures at the wavelengths of about 1 to 100 meters which were supposedly completely attenuated. No value is given for the reduction of any signals, especially those naturally coming from the sender as the "blackbody" radiation described earlier.

In summary, we can say that Vasiliev's results are inconclusive as to the reduction of telepathic signals by shielding. There are other experiments which attempt to investigate the same question, but these also seem to be vulnerable to the same criticism as Vasiliev's work. Because of this we will have to leave open the possibility that telepathy can be achieved, if at all, by radio waves of a wide range of frequencies and wavelengths. We must look at other aspects of telepathy to see if the phenomenon can fit in with electromagnetism.

A series of investigations into this question was performed beginning in 1962 in the dream laboratory at the Maimonides Medical Center in New York. The tests occurred using a variety of senders and receivers, with various distances between the transmitter and receiver. The dreams of the receivers were described in the laboratory on the following morning and matched to a collection of pictures, including the one actually concentrated on by the sender. The conclusions of the series of tests (which were partially replicated elsewhere) were that:

a. Telepathy in dreams can be demonstrated in a laboratory setting.

b. The elements of orientation, expectancy and volition appear to be necessary for extrasensory effects to occur in dreams.

c. Male subjects have been more effective telepathically in dream experiments as receivers than female subjects.

d. Telepathic effects have occurred in dream experiments when subject and agent have been separated by ninety-six feet, fourteen miles and forty-five miles (though they proved unsuccessful over a distance of two thousand miles).

e. Target stimuli that are emotional in nature appear to be more effective than nonemotional material in the dream experiments.

f. Two initial dream studies have supported the precognitive hypothesis.

g. Altered states of consciousness appear to be favorable for the occurrence of ESP.

The above items, except f., are all relevant to the questions about telepathy which were raised earlier. Item e. agrees with the work of Thelma Moss mentioned above. From item d. we see that an indefinite extension of telepathy to very long distances (thousands of miles) may not occur. To make this consistent with the spontaneous phenomena recounted earlier, we have to realize that in these latter the sender of the telepathic signal is in a far more "excited" state than is an agent in a telepathy experiment. If we very naturally assume that the agent has to expend some form of energy in his transmission, it is to be expected that at a time of personal crisis the sender might well call on normally inaccessible reserves of energy. Spontaneous cases are therefore expected to generate telepathic signals over far longer distances than would be the case in more controlled laboratory tests.

The many anecdotal cases reported on the use of telepathy have been backed up by what would appear to be careful laboratory investigation, although these do have some difficulties associated with them.

A detailed model of telepathy has been suggested by the Russian scientist, I. M. Kogan. In the 1960s he attempted to see if the idea that telepathy is transmitted by means of a low frequency radio wave fitted the data. He related the amount of information that could be carried by such a radio wave to its strength and its range of frequency. Radio propagation theory was also used to find out how much current flow in the sender's body must occur to transmit a given amount of information a given distance. Most interestingly, Kogan then presented data which, he claimed, showed a falloff of telepathic transmission with distance explicable only in terms of radio waves of long enough wavelength or low enough frequency (where frequency and wavelength are inversely proportional to each other).

The fact that there exists an upper limit to the wavelength arises because otherwise too large a current would be required in the sender's body or brain (compared to what is usually measured) in order to produce an effective signal. There also exists a lower limit to the wavelength of the radio wave, since otherwise it could not be produced by natural brain activity nor travel long enough distances over the earth's surface. When all this is put together, Kogan claims that there results a wavelength in the range of three hundred to one thousand kilometers, or frequency in the range of three hundred to one thousand cycles per second. As he writes at the conclusion of his paper: "There are reasons for assuming that the existence of telepathy does not contradict the laws of nature, and its carrier could be the electromagnetic field of extra-long waves excited by biocurrents." However, further calculations show that the electric power that would be induced in a receiver by radio waves of such wavelength as Kogan accepts is at least a million times too small to be noticeable in the typical telepathic situations considered by him. This also applies to the card-guessing and other telepathic tests, and to the anecdotal reports of telepathy over long distances that are so dramatic.

Thus radio waves will not explain telepathy either. We have therefore reached an impasse; distant viewing and telepathy have strong evidence in their support but definitely contradict modern scientific understanding. How can we reconcile these contradictory features? The next step would be to go back to the evidence for these phenomena and look at it even more carefully. Is it, in fact, as good as has been claimed by its proponents?

We could consider the evidence piece by piece to point out flaws of this or that sort. The SRI results described in the previous chapter depend heavily on matching the scene described and drawn by a remote viewer with one of a set of possible places. This process of judging is clearly subjective, since for one of the subjects, Pat Price, six judges assigned scores to his results of 7, 7, 6, 5, 3 and 3, respectively, instead of the 1 expected by chance guesswork. Due to lack of consistency among the scores, it is clear that there are no precise criteria specified to indicate what is to be regarded as a successful vision of a distant place.

It could well be that there is a fault in some other part of the test, and in the other tests mentioned as successful in this or the previous chapter. There is also the question of the level of significance. Are odds of a hundred to one against chance the correct level at which results are to be regarded with respect? Or should the odds be a million to one? These and numerous other questions will have to be answered in detail before the data on distant viewing can be accepted for more than tentative indications of something requiring further investigation.

There remains the possibility of a supersensory mechanism in the body which could detect magnetic field variations on the earth's surface. We have therefore to try and see if such a mechanism exists.

Various attempts have been made to find such a "supersensory" mechanism. The most reasonable of these have to do with dowsers (those who claim to be able to detect the presence of water or metals by using a forked twig or similar means). It is said that dowsers sense local alterations in the magnetic field of the earth brought about by the presence of buried iron objects or underground water. Such sources will produce electrical potential differences between distant points, which will generate minute electric currents. These in turn will generate small magnetic field charges which, it has been claimed, in particular by the French physicist Y. Rocard and the American engineer Zaboj Harvalik, are detected by the experienced dowser.

Rocard used artificial magnetic fields created by passing an electric current through a large coil of wire. The dowser walked past the coil and had to detect if the current was on or off. A preliminary trial run was allowed in which the dowser was told whether the current was on or off; the following five attempts were then 100 percent successful. No success was obtained if the trial run was not given.

A like trial was also attempted by the British Ministry of Defence (and is discussed in more detail in Chapter 10). A setup similar to that used by Rocard was constructed and an experienced dowser was asked to see if he could detect if the current was on or not. Seventy-five trials were attempted, but only chance level of success was obtained. Thus Rocard's results were not repeatable.

The work of Harvalik achieved a completely different level of sensitivity and also used a different test situation. The main emphasis was to detect the sensitivity of people, using the dowsing reaction, to a varying magnetic field. This latter was set up by creating a controlled flow of electric current through the earth from one metal pipe to another some sixty feet away. The dowsing subject walked across the line of flow and the strength of the dowsing reaction which resulted as he crossed the line was noted. As Harvalik wrote: "He [the subject] starts his walk about twenty feet from the midway point, crosses it and continues to walk another twenty feet beyond the midway point. If a current of sufficient intensity passes through the ground the dowser would show a signal when he approached the midway point, about five feet from it. The signal would disappear about five feet beyond the midway point."

The detailed results reported by Harvalik, in a long series of papers, are quite astounding. There are essentially two different features. One is that many people appeared to be able to show a dowsing reaction to a very low level of magnetic field. The value of this was about five hundred thousand times less than that naturally occurring on the earth's surface, which is about half a gauss. Such small magnetic fields are usually measured in a unit called a gamma and denoted by the appropriate Greek letter γ, one γ being a hundred thousandth of a gauss. The sensitivity discovered by Harvalik in a good proportion of the two hundred people he had tested was thus about a tenth of a gamma (and could be even less). In other words, people are sensitive to a millionth of a gauss. Another test on fifty-four males was claimed to show that eleven did not react at all, while thirty-four were sensitive to about a tenth of a gamma, the remaining nine being even more sensitive, one even to a thousandth of a gamma.

The second, and more surprising result, was obtained in work with W. de Boer, a professional dowser from Bremen in West Germany. The same test setup was used and came up with the microscopically low figure of about a hundred thousandth of a gamma, or about a hundred times more sensitive than the most sensitive male in the tests described above.

These results are, indeed, fantastic: over half the population sensitive to magnetic fields about a *half a million times* weaker than that of the earth; one person even a hundred times more sensitive. If this were true it would be a disturbing result, since there are fluctuations of the earth's field of many times a gamma every day which could affect many people in some way. Even the magnetic fields around electric lights would produce much larger fields. Yet there are no well-established reports of people being sensitive to such features of the environment.

A dowser was tested by E. Balanovski and myself to see whether he could detect the presence or absence of a magnetic field of 500 gauss. He said that he had absolutely no signal at all from it, and refused to continue any further, remarking that he did not wish to waste our time.

I also tested to see if *I* could sense the presence of a reasonably strong pulsed electric field (pulsed up to a million cycles a second and about a hundred to a thousand times stronger than the earth's static electric field). This I did, again with my colleague Balanovski, by seeing if my speed of reaction was altered when the field was on my head and shoulders. At no time did I find any effect, either on myself or on Balanovski.

From this limited set of results I can only surmise that there is little likelihood of the existence of a magnetic sensor in man, in contradiction to Harvalik's results.

7
Precognition

THE PREMONITION OF FUTURE DISASTER HAS A LONG HISTORY.
There are many cases where precognition was used to determine
future action, and famous oracles, such as that at Delphi, were
often consulted. The occurrence of precognition in dreams is
also described in a number of places in the Bible. Indeed eighteen of the thirty-nine books of the Old Testament are headed
"The Book of the Prophet." One of the most quoted precognitive dreams of all times is that of the Pharaoh which was interpreted by Joseph. In the New Testament, the day after the birth of
Jesus his earthly father was warned in a dream to flee to Egypt
to avoid the wrath of King Herod. There seems never to have
been a dearth of premonitions of disasters.

In order to begin our analysis of precognition we first have to
distinguish it from prediction. This latter has been brought to
great precision by the aid of the scientific method. For example,
the times of the eclipses of the sun can be worked out for the next
millennia with high accuracy. Predictions are also made about
human affairs and these may turn out to be hopelessly wrong.
Niels Bohr, the founder of nuclear physics, predicted, only a
decade before the first atomic bomb explosion, that the power of
the nucleus would never be tapped. The head of the British Post
Office, on hearing of the invention of the telephone in America,

is reported to have remarked, "We will have no need of such a thing; we have plenty of messenger boys." Prediction may be proved remarkably correct, as it usually is in the hard sciences, or hopelessly wrong, as when human affairs are involved. But failure is expected because the process essentially involves educated guesswork.

An occurrence that used to trouble me frequently was a premonition, as I sat working at my desk, that my telephone was about to ring. I would reach out my hand to pick it up and sure enough a few seconds later it would actually do so. The experience can be unnerving, especially if it happens several times in one day, as has been the case. I finally tracked down the cause of this phenomenon—I found that a brief warning "click" occurred as the telephone was actuated a few seconds before the bell actually rang. I noticed this click only after my habit of picking up the telephone just before it rang had occurred too many times to be purely by chance, making me alert to just such an audible cue.

Precognition involves knowledge of future events which cannot be inferred from present knowledge, and so is very different from prediction. Prediction of the time of a future eclipse of the sun is only a more precise version of Jeane Dixon's supposed "precognition" in 1956 that "the 1960 election . . . will be won by a Democrat but he will be assassinated or die in office." This "precognition" was dramatically proved on November 23, 1963, when President Kennedy was assassinated. Jeane Dixon could have based her prediction on a past history of presidential assassinations having taken place not infrequently in the past. Precognitive dreams, "visions," insights, etc., were unnecessary. There may even be a regular pattern of such assassinations, so a projection of such a pattern into the future would make such a prediction more accurate.

A typical case of precognition was reported to me recently. It involved a woman who awoke in the middle of the night having just experienced a very unpleasant nightmare. In it she was trapped in a tube filled with smoke and very poorly lit, though she could make out what seemed to be jagged edges. She also heard people screaming and sobbing. The woman woke her husband and told him of the nightmare. The following day the Moorgate tube disaster occurred, one of the worst underground train accidents in the history of London. In it an underground train ran past the platform and the front carriage was crushed into the dead end of a tunnel. Survivors described the scene in very similar words to those used by the woman about her nightmare. She telephoned me and was clearly troubled by the incident.

There are several premonitions bureaus which have been set up in various cities to record dreams people have had and check them against future disasters. Among them are the Central Premonitions Registry in New York City and Premonitions Registry Bureau in Berkeley, California. Some of these premonitions are indeed quite remarkable, at the same sort of level as the Moorgate tube disaster dream described above.

Instances of similar premonitions of the same disaster have also been reported. After the Aberfan mine disaster in Wales, in which a disused mine tipple collapsed and killed over a hundred people, reports were collected from a number of individuals who claimed to have had premonitions of the accident. Of the seventy-six reports received, twenty-four premonitions had been witnessed by another person before the event.

This and evidence of a similar kind has led many psychic researchers to accept the existence of precognition. Jeffrey Mishlove, in his very extensive book on all aspects of the paranormal, *The Roots of Consciousness,* says, "There is a good deal of evidence to warrant that precognition actually does occur—with all of the ramifications regarding time and free will." Another well-known psychic researcher, E. Douglas Dean, has written recently, "Hundreds of studies have been made of spontaneous precognition experiences and of controlled experiments in the laboratory. The use of modern electronic instruments has confirmed this. Even dreams have been shown to come true in the laboratory."

We described in Chapter 6 some of the attempts made at the Maimonides Dream Laboratory to discover the level of telepathic ability during dreaming. Tests were made in these conditions to see if a subject would incorporate in his dreams of a given night an experience which he was to have the following morning. The experience was built around a randomly selected word; for example, when the word "teaspoon" was selected the previous day he was given soup to eat with a teaspoon the following morning. The success or not of the recounted dreams over eight nights was judged by three independent judges, and found to be explicable as happening by coincidence less than once in a thousand such tests. In a study lasting a further sixteen nights, the subject was exposed on alternate days to the packaged experience (slides and tape recordings) the evening following his attempt to precognize it. The dreams on the pre-experience nights were then compared to those on the post-experience nights. Judges again concluded that there was a significant relation between the pre-experience dreams and the experience but not between the latter and the post-experience dreams. The odds that the former relations could have been purely chance were less than one in a thousand.

The difficulty in accepting these results as strong evidence for precognition is that the judging itself is not as reliable as one would like. The process of assessing to which one of a set of pictures a given picture is closest is subjective, different judges giving different conclusions in various cases. How many judges should be chosen before a reliable assessment could be expected? If they are all using a set of criteria incorrect in some fashion, then however many judges are used the result will still be wrong. When written records are added to the pictures, the problem becomes even more difficult. Sets of different criteria should be chosen to see what gives the most accurate method in known cases. The ideal method would be to have a computer program designed to incorporate all reasonable criteria. Until that is done, odds of one in a thousand are meaningless.

An attempt was also made to see if hypnosis accentuated precognitive abilities. Under hypnosis the subject guessed the sequence of numbers which was to be generated the next day by throwing a ten-sided die to look up entries in a table of random numbers. The subject was also asked to say, after a guess, if he thought it was reliable. Out of 1,950 numbers a particular subject guessed 175 of them correctly, just about the number expected by chance. However, the subject got 60 right out of 360 when he thought he was reliable and only 115 right out of 1,590 when he did not. In other words he was successful 16.7 percent of the times he expected to be but only 7.2 percent of the times when he did not. This difference does indeed appear considerable, as it would be expected to occur by chance only twice in a hundred million such tests.

The most remarkable of all results on precognition has been obtained by using a random number generator built specially for the purpose of ESP research. It was designed by Helmut Schmidt, one-time director of Rhine's parapsychology laboratory. The generator works by means of an interrupted current with a million pulses a second. A switch in the circuit is closed by a subject at a certain time. The presence or absence of a pulse at that time is used to light a lamp or leave it dark. To prevent any periodicity of switching being used by a subject, a random time delay is interposed between the time the switch is pressed and the presence or absence of the current pulse. This time delay is produced by the radioactive decay of strontium-90 nuclei.

Precognition is tested with this machine by a subject guessing beforehand whether the lamp will be dark or lit. It was adapted so as to randomly light up one of four bulbs, the subject's task being to guess beforehand which of these bulbs would be the one. The results of this test were astonishing. Two subjects attempted to guess better than average and did so 189 times more than expected by chance in just over ten thousand such attempts. One of the two subjects, along with a new one, also attempted to guess less than the expected number of successes; they did so 212 times less than would be expected by chance in a total of just under ten thousand trials. The odds against this happening by chance are astronomical—more than ten billion to one. Other tests of a similar nature gave slightly smaller odds of about a billion to one.

The machine was tested many times before, during and after the tests and always found to generate a good sequence of random numbers, at least as far as the frequencies of single numbers and blocks of numbers were concerned. The data were recorded automatically, so that human recording errors were completely excluded. These tests seem to be immune to the whole gamut of criticism leveled against the card-guessing tests. Do they finally validate one of the most puzzling psychic phenomena of all: true precognition?

We should first remark that tests for randomness must be done with great care when sequences of thousands of numbers are involved. More careful tests than checking frequencies of single numbers and pairs may be necessary to make sure that a truly random series is indeed being generated at the required level of accuracy. The next feature we should notice is that if the four trials were put together, the success rate above the chance value of 5,000 in the total of 20,000 tests is only 23, completely insignificant. Are the results not an example of how selected parts of a set of guesses can be more or less successful than chance, even though the total set of successes is only at the pure chance level? Only further trials would indicate that, and clearly these need to be performed. Finally, we must realize that only a few machine errors would be needed to achieve the results. Machine checks at the highest level would clearly be necessary.

The protocol used by the Stanford Research Institute investigators, Puthoff and Targ, for their tests of precognition merely required the distant viewer to describe the scene fifteen minutes or so before the test team went there and even before the destination was known by the team. This latter question was decided by generating a random number between 0 and 9 and choosing the test site associated with the corresponding number.

The results obtained by one subject in four separate tests were reported as very good. The second target visited in the tests was the fountain at one end of a large formal garden at Stanford University Hospital. The subject described a formal garden behind a wall with a "double colonnade" and "very well manicured." According to the report on this work, published in an engineering journal in 1976, Puthoff and Targ wrote, "When we later took the subject to the location she was herself taken aback to find the double colonnaded wall leading into the garden just as described." In this and the three other tests, her descriptions would have fitted the target locations by chance only once in twenty-five times, according to the assessment of three judges. The investigators seemed to regard this as a good validation of precognition. They state: "For reasons we do not as yet understand the four transcripts generated in the precognition experiments show exceptional coherence and accuracy as evidenced by the fact that all of the judges were able to match successfully all of the transcripts to the corresponding target locations."

It is necessary to consider the report written by Puthoff and Targ somewhat further to assess the level of their evidence. We have already made the criticism that the judging procedure itself is of dubious validity. Nor is it clear what odds against chance should be regarded as significant. But the level of expertise behind the report becomes clear when the two scientists turn to analyze models of distant viewing and precognition.

After remarking that "we have no precise model of this spatial and temporal remote-viewing phenomenon," they proceed to state: "It is important to note that many contemporary physicists are of the view that the phenomena we have been discussing are not at all inconsistent with the framework of physics as currently understood." They present Kogan's arguments and others for some radio-wave type of propagation, though they indicate the need to extend this approach of "radio-wave" distant viewing to the precognitive case. This, they suggest, can be done by what are called "advanced" waves, which violate causality—the principle that cause always precedes effect. For an "advanced" wave, effect always precedes cause. For the feasibility of such a bizarre feature of advanced waves, they appeal solely to a sentence in a textbook on electromagnetism that states (correctly) that advanced waves are logically possible in physics and hence should only be discarded on empirical grounds. Precognition might indicate that such a neglect is unjustified.

To consider this question more closely, let us note first that it has already been shown in Chapter 5 that the radio-wave explanation of distant viewing fails because the distant objects "observed" are too small to have been noticed by appropriate radio waves. Such an analysis does not appear to have been performed by Puthoff and Targ, nor do they seem to take seriously their own results that shielding the subjects from radio waves does not reduce the success rate of the tests. Thus, while the usual "causality-satisfying" radio waves (called "retarded" waves) certainly should be ruled out as explaining precognition, the causality-violating advanced radio waves are not, according to Puthoff and Targ, supposed to suffer the same fate. No argument is given that these advanced waves are any better than the retarded waves in making fuzzy objects look clearer, nor that the advanced radio waves can penetrate the shielding impenetrable to the retarded or normal radio waves.

There is an even greater difficulty than the theoretical ones raised above, and that is the violation of causality which precognition brings about. Before we consider that, we must reiterate the difference between prediction and precognition. If the knowledge were given of where every particle was at a given time, and with what velocity, then predictions as to the state of the world at any time in the future might be possible. This would be so, at least in principle, in a purely materialistic world in which the motions of material determine everything, including mental attributes. In terms of such a full-blown materialistic approach, it would be possible to have complete knowledge of the future. Precognition would thus have been superseded. However, even if we do inhabit a completely materialistic world, we still do not possess enough knowledge to have such an embracing view of the future. Precognition is that ability which would allow us to know about the future without enough knowledge to predict it with any certainty. There may still be partial knowledge, allowing for a certain degree of confidence in one's future. Precognition claims to give knowledge of the future with a certainty far exceeding such predictions.

Let us turn now to consider how such information might be able to pass from the future to here and now. What would seem to be required is a mechanism for violating causality, the supposition well-based on experience that, as we mentioned above, cause always precedes effect. Thus, if a boy throws a stone at a window, the glass shatters after it has been hit, not before. Precognition would seem to require the effect to exist, in some fashion, in advance of the cause. Therefore a person sitting on the other side of the window may precognize its being shattered and so avoid being seriously injured by flying glass. He would do so by sensing, through his precognitive ability, the glass breaking and showering over him. For example, he might have a visual image of that occurring. The real effect, the flying glass, would have affected his brain cells in some way before the cause (the stone) was thrown by the boy at the window.

Signals of some form must arise from an effect to produce changes in the environment at an earlier time than the cause of the effect. In other words, the effect must itself become the cause of effects occurring before the cause of the original effect. One cannot avoid this situation by requiring these earlier effects produced by the later one to be "mental" ones only. For sooner or later there must be physical activity produced by the later effect, and unless this activity occurs before the original cause, the phenomenon could not objectively be called precognition. There must be at least a statement of the precognition to some sort of witness.

Violation of causality has been investigated very carefully in physics. It is clear that the extremely precise results obtained over the past few decades have been completely ignored (very likely because they are unknown) by those working in parapsychology, especially those involved in precognition. It could well be that these results are not well known because they are esoteric, being understood by the rather small fraternity of high energy physicists. The results have been couched in language inaccessible to most. But it should not have been so to physicists like Puthoff and Targ. Since the point is basically simple to explain, I shall attempt to do so now.

When a beam of fast-moving elementary particles are shot at a target, the latter will recoil and the particles will be scattered away from their original direction. A very sensitive test of causality is to determine if there is any recoil of the target a short time before the particles of the beam hit it. It is possible to measure down to extremely short time intervals in such a process by using all the data on the scattering at all energies up to the highest available. This data is then fed into a "dispersion relation"—so-called because it gives the amount of spread or dispersion that has taken place in the particle beam. If causality is valid the data should correspond with the dispersion relation. Any discrepancy could only be explained as produced by a violation of causality.

In this way, violations of causality have been probed down to distances less than the natural sizes of the elementary particles themselves, about 10^{-14} cm. None have been discovered. The corresponding energies of the projectiles are billions of times above chemical energies. If causality violation were to be the modus operandi of precognition, it could not be possible to do so by means of the usual energies available in the human body. Anyone who attempts to claim the opposite is contradicting very well-verified facts of nature.

Let us return, then, to the claims made by Puthoff and Targ and other parapsychologists that physics itself allows violations of causality, especially in terms of the "advanced" waves mentioned earlier. This approach has in fact only been proposed in physics in cosmological models, on which there is extensive literature. These applications of advanced waves to cosmology were found to have great difficulties in them and have now been abandoned. In any event, all such approaches involved attempts to remove any observable causality violations and so agree with the elementary particle results described above. Furthermore, there is no relation at all between the advanced waves and human consciousness in these theories. That such does occur has been claimed solely by the parapsychologists, and then only with data from precognition to go on, so they have no firm physical basis at all on which to found such an approach.

Much has also been made of Dirac's theory of the positron, the positively charged companion to the electron, and Feynman's reinterpretation of it. This does contain some apparently bizarre features, but nothing along the lines, claimed by the parapsychologists, of particles going "backward in time." Let me describe the situation. In 1929 Dirac proposed a theory of electrons which embarrassingly also described particles with negative energy. The absence of such a negative-energy particle in a completely filled "sea" of such particles would actually be observed as the presence of a particle with positive energy and positive charge. This suggested positive particle (the positron) was discovered by Anderson in cosmic ray tracks in 1932. Later, Feynman showed how the original hypothetical negative-energy particles went "backward in time." Since we only observe particles of positive energy, he showed how it was possible to reinterpret these postulated negative-energy, backward-going particles as positive-energy, forward-going particles. Since only positive energies are possible, particles going backward in time cannot exist. Any claim that they do is purely a fantasy in the mind of a parapsychologist.

There is therefore no direct justification for precognition from physics. It would seem that experimental evidence from high energy physics is strongly against it. The claim noted earlier that "many contemporary physicists are of the view that . . ." is thus completely false.

There is one way in which precognition might be explicable. That would be in a world in which materialism is not correct, and the direct action of mind on matter were possible. Violations of causality might then arise in some way in such a mind/body interaction. We discussed in Chapter 3 the importance of materialism as the only basis for precise explanation. Its rejection can only lead at best to silence, at worst to irrationality.

It might be claimed that it does not matter if parapsychologists know nothing about fundamental physics as long as their tests of various paranormal phenomena are valid. Yet an understanding of modern physics would make them, and others, treat the parapsychologists' odds against chance and other parts of their tests with far greater care.

Two separate colleagues of mine have tried to observe the tests at SRI, but have been refused. Since they are both sympathetic to the psychic cause and had carried out tests of their own—one on dowsing and the other on spoon bending—their exclusion seemed surprising. It may well be impossible to learn exactly how the tests were done. From the above remarks I can only conclude that the results of the distant-viewing and precognition tests cannot be believed.

I said previously that after looking at the earlier evidence for paranormal phenomena it would be necessary to conduct tests oneself. This clearly seems to be such a situation. Yet the attempts I have made to validate distant viewing, telepathy or precognition have all ended in dismal failure. For example, Eduardo Balanovski and I tried a preliminary trial of distant viewing with the English psychic Matthew Manning. It proved completely unsuccessful. I also tried to validate the claims for precognitive powers among business executives that had been put forward by American parapsychologists; again I failed.

We can summarize the argument so far, regarding distant viewing, telepathy and precognition. Anecdotal cases are most probably the results of coincidences, which do indeed happen. The tests which have been claimed to give support to the existence of these phenomena are very likely based on the use of shaky statistical analysis of a poorly designed experiment. Attempts to duplicate certain of these tests have completely failed to do so. Finally, theoretical arguments put forward by the parapsychologists for the feasibility of such phenomena have been shown to be based on distortions of the writings of theoretical physicists as well as complete ignorance of highly relevant and important areas of physics.

8
Psychokinesis

OF ALL SEEMINGLY PARANORMAL PHENOMENA, THE ONE MOST obvious in its effects is undoubtedly the poltergeist or "noisy ghost," so-called because of the raps and knocks that are produced in many such cases.

The commonsense reaction to events such as those reported at the lawyer's office in Rosenheim (described in Chapter 2) is to say categorically that they must be caused by fraud, as most of their different features contradict normal experience. A poll taken in West Germany during the sixties by the Freiburg Institute for Border Areas of Psychology and an institute for public opinion research found widespread scepticism among the adult population: 72 percent of those polled attributed the phenomena of poltergeists to pure superstition, 10 percent were unsure and only 18 percent were convinced of their existence.

That there have always been believers and disbelievers is shown by a case from 1575 on file at the Supreme Court of Paris. The tenants of a house claimed that they were being disturbed by a poltergeist and wanted their tenancy agreement annulled. The lawyer for the landlord urged that it was a shame to believe in the nursery tale of poltergeists and that the gullibility of uneducated people should not be encouraged. In 1660 Joseph Glanvil, one of the first members of the British Academy, began a

report of a poltergeist case by saying that he knew very well
". . . that the present-day world treats all such stories with laugh-
ter and derision and is firmly convinced that they should be
scorned as a waste of time and old wives' tales. . . ."

In our search for the truth we must avoid such extremes, keeping our eyes open but our critical faculties alert. Whatever the
evidence of fraud or self-delusion, we may also discover valid
poltergeist cases.

Reports investigated by the French police between 1925 and
1950 have been collected and analyzed by a French police officer,
E. Tizane. He discovered that there was a classification of the
phenomena in increasing order of strangeness:

a. Bombardment. Often a house becomes the object of a real
hail of projectiles. Stones fall on the roof, break panes and
penetrate through openings. Phenomena rarely occur in the
interior of the house once outside bombardment from the
exterior begins.

b. Bangs against doors, walls or the furniture are heard, sometimes at the same place and sometimes in all parts of the
house.

c. Doors, windows and even securely closed cupboards open
by themselves.

d. Objects are skillfully moved or thrown. Fragile ones are
often unbroken, even after a move of several feet, while solid
ones are sometimes completely destroyed.

e. Bizarre cracks and noises are sometimes observed.

f. Displaced objects sometimes do not show a "normal" trajectory. They behave as if they have been transported and may
even follow the contours of furniture.

g. In some rare instances, foreign objects penetrate into a
closed space.

h. When handled by observers, the objects give a sensation of
being warm.

i. Objects seem to form themselves in the air.

There are indeed many cases which fit into one or another of these categories and which on first sight seem reasonably well authenticated. Thus a case was reported in Aberdeen, Scotland, in April 1973 involving two elderly ladies of seventy-two and seventy-five years of age and the ten-year-old granddaughter of one of them. They all lived together in a typical residential house of that area. In the early part of 1973 repeated heavy knocking and furniture movement continually disturbed the elderly ladies' sleep. These noises and movements were investigated by the president of the Aberdeen Psychic Society and his wife at the instigation of the police, to whom the elderly ladies had appealed. The son of one of the ladies also tried to find the source of the disturbing effects. These were reported as including: "Movement of dining room chairs. Seats separated from chairs and either or both clattering downstairs from the room to the front vestibule. Coffee table turned upside down, then thrown downstairs. Opening and closing of doors. Light switches manipulated on and off. Cushions, ornaments, shoes and clothing moved from one room to another. Heavy rappings on floor boards. Small mats flying about and hitting doors. Telephone falling off table (three times in one evening)."

The Aberdeen Psychic Society investigators spent a night in the affected house. The report says: "Within five minutes of the child going to bed a tremendous knocking and thumping took place outside the door of the room we were in; . . . when I first opened the door the small mats were in motion—they were being whipped about against the door, but as soon as we went out to the landing they dropped to the floor." The young girl was suspected as being the source of the effects, yet she could not easily have directly caused the movement of the mats in the hall since it was stated that she was in her bedroom at the time.

The girl's uncle stayed in the house for five nights and observed a wide range of poltergeist effects. His report stated that among many bizarre effects he saw were "lights put on and off. Switching actually seen in movement. [Switches are quite stiff.]" The uncle was certain that the girl was at the center of the phenomena, and he observed: ". . . they started while the child was beginning to feel drowsy. It continued while she lay in bed awake, providing nobody was with her or taking up her attention, and became very strong when she was in the first stages of sleep. It stopped when she went into deep sleep. When taken unawares, sitting up in bed during the phenomena, she seemed quite unafraid and enjoying all that took place around her."

The following day the child was taken to a children's home, where it was felt a more suitable atmosphere could be provided for her. No unusual things occurred around her from then on, nor in the house she had just left. It is reasonable to assume that the home atmosphere, which was associated with the death of the girl's divorced mother from cancer nine months before and of her grandfather three months earlier, caused the girl to be severely disturbed emotionally.

There are many cases of a similar level of proof, each with its testimonies from credible witnesses who appear not to have observed any fraud in association with the abnormal movements or noises. However, it is not only the *credibility* of the witnesses that is crucial but their acuteness of observation. For example, there is a class of poltergeist phenomena which can very likely be ascribed to the subsidence of the foundations of the house in which they are occurring. This explanation was put forward by W. G. Lambert in 1955 with special relation to effects caused by underground water moving in subterranean rivers or channels. Certain famous poltergeist cases can be explained in this way, and it has recently been proposed that exactly this was the cause of the movements and noises in London supposedly achieved by the great English medium, D. D. Home, from 1855 to the late 1880s.

Acute observation by the witnesses is also essential in order to rule out fraud. This is especially important in poltergeist cases since these have nearly always involved a person (usually a child) under great emotional stress and possibly prepared to go to considerable lengths to draw attention to him/herself. Such tension was present in the girl at the center of the Aberdeen case mentioned above. A similar problem could have arisen in the Sauchie case reported by the Canadian psychic researcher, A. R. G. Owen.

Sauchie is part of a sprawling town at the limit of navigation of the River Forth in Scotland. Abundant poltergeist activity occurred there during the last two months of 1960, being associated with a girl of eleven. Five witnesses of standing in the community, the minister of the local church, three medical doctors and the girl's teacher, observed some of the phenomena firsthand. The events started one evening with what sounded like a bouncing ball heard in the girl's bedroom. The noise then traveled down the stairs and into the living room with her. The next evening knocks loud enough to be audible over the whole house were heard by the family and neighbors. The minister was called and arrived at about midnight. He found the knocking to come from the head of the girl's bed even though the girl could not push or strike the headboard in any way. Nor could the noises have come from vibration in the wall since the bed was not touching it. The minister held the headboard and felt it vibrating in unison with the noises. He also saw inexplicable movement of a large linen chest over two feet long, about one and a half feet high and over a foot wide, and full of bed linen. It was beside the bed, and started to rock sideways before moving along its length with a jerky motion for a distance of about eighteen inches, and then returning. The box was so heavy (at least fifty pounds) that it took two men to lift it out onto the landing.

The next evening further noises were heard, as well as some movement. The following day at school her teacher noticed the lid of her desk raising itself. Another strange phenomenon was observed by the teacher at the beginning of the following week. The girl had come to stand beside the teacher's desk (actually a table, four feet by two feet, quite solidly built) to ask for help on a problem of arithmetic. Suddenly the teacher noticed the blackboard pointer lying on her desk beginning to vibrate and move across the desk so that it eventually fell to the floor. As this movement was occurring the teacher felt a vibration in the desk when she placed her hand flat on it. The desk also began to move till the teacher was no longer sitting at the middle of one side nor near enough to be able to use the tabletop to write on comfortably. The teacher then looked at the girl and found her still standing with her hands clasped behind her; she started to cry and said, "Please, Miss, I'm not trying it."

On the following evening two doctors visited and heard

". . . several outbreaks of 'knocking.' These varied from gentle tappings to violent agitated raps. . . ." They were satisfied that ". . . the sounds came from within the room but were not due to the activity of anyone inside the room." Two evenings later they were joined by the girl's doctor, who had already visited and heard knockings and seen the linen chest move. At that time he had heard the knocks when the girl was lying on the bed with the bedclothes removed to prevent her using covert movement of her body or limbs to produce the noise. He observed waves or "ripples" passing over the bedclothes and an inexplicable 90-degree rotation of the girl's pillow; he also saw the linen chest move and its lid open and close several times. When all three doctors were present there was a steady stream of tappings (which were recorded on tape) and occasional puckering movements of the bedcovers (though sadly they failed to film these motions with a movie camera due to the brevity of the occurrences).

The various observers each carefully considered the possibility of fraud and was able to exlude it to his or her own satisfaction. Thus the minister saw the linen chest move when the girl was lying flat in bed and well tucked in and no one else near the chest; one of the doctors saw movement under conditions equally good enough to exclude fraud; and the three doctors and the minister were all satisfied that the noises could not have been produced by the girl by trickery nor by an accomplice (possibly one of the other two children in the family) outside the bedroom. The teacher was careful to note whether the girl could physically have caused the movements of the two desks and concluded quite definitely that she could not have done so.

The explanation that the effects were caused by the sudden flow of underground water was ruled out by the local Surveyor's Department, which stated that though there had been old workings in the vicinity the overlying ground was stable, with no subsidence, vibration or ground movement. Nor did the house show any signs of cracking as would be expected from differential movement of the ground. No underground streams were known in the vicinity.

The disturbances apparently tailed off by the early part of

1961, and while this case is of great interest it was not investigated by someone with previous experience of such phenomena. This disadvantage in comparison to the Aberdeen case mentioned earlier is somewhat outweighed by the greater amount of carefully witnessed activity.

Both valuable features—that of considerable poltergeist activity and its extensive investigation—are present in several cases looked at by another experienced psychic researcher, W. G. Roll of the Institute of Parapsychology at Chapel Hill, North Carolina, in the U.S.A. One of the best of these seems to be the disturbances which took place in the Herrmann household in Seaford, Long Island, in the middle of February 1958.

The household consisted of Mr. and Mrs. Herrmann and their son and daughter who were then twelve and thirteen years old, respectively. Noises were heard and small objects in the house, such as bottles, were caused to move in such an inexplicable manner that the father even lodged a complaint with the local police department. There was one event in particular which precipitated the father's call for outside help. As he said: "At about 10:30 A.M. I was standing in the doorway of the bathroom. All of a sudden two bottles which had been placed on the top of the vanity table were seen to move. One moved straight ahead slowly, while the second spun to the right for a 45-degree angle. The first one fell into the sink. The second one crashed to the floor. Both bottles moved at the same time." The father noticed that his son had frozen in his tracks when the bottles began to move. The police record of the case adds further to this event: "Mr. Herrmann standing in the bathroom doorway, his son James at the sink brushing his teeth, actually saw a bottle of Kaopectate move along the formica top of the drain in a southerly direction for about eighteen inches and fall into the sink. At the same time a bottle of shampoo moved along the formica drain in a westerly direction and fell to the floor. There was no noise or vibration and no one touched either bottle to move them."

The police assigned a detective to investigate the case, and a little later Roll and another experienced psychical researcher, J. G. Pratt, visited the house for a week. In all, sixty-seven different movements of objects were noticed during the poltergeist activity. There were three particularly interesting types of events: those when someone saw the movement of an object with no physical cause (four cases), occurrence when there was no one near enough to have caused them (thirteen cases) and movements or noises which occurred when the two investigators were in the house (five cases). Of these latter, two were noises and the other three were inexplicable movements of objects.

One of these latter cases arose when all present in the house were downstairs: Mr. Herrmann was telephoning in the dining room where his son was sitting at the table, Mrs. Herrmann and the daughter were in the kitchen preparing dinner, and a police sergeant and the two investigators were also downstairs. A noise was heard from upstairs and it was discovered to have been caused by a lamp having inexplicably overturned and fallen on a bottle.

That this was not an isolated case of strange noises and movements is shown by the report taken from the police records: "At about 10:15 the whole family was in the dining room of the house. Noises were heard to come from different rooms and on checking it was found that a holy-water bottle on the dresser in the master bedroom had opened and spilled, a new bottle of toilet water on the other dresser in the master bedroom had fallen, lost its screw cap and also a rubber stopper, and the contents were spilled. At the same time a bottle of shampoo and a bottle of Kaopectate in the bathroom had lost their caps and fallen over and were spilling their contents. The starch in the kitchen was also opened and spilled again and a pot of paint thinner in the cellar had opened, fallen and was spilling its contents."

A middle-aged cousin visited the house during this period and reported that as she was seated in the living room across from the boy, who was in the middle of the couch with his arms folded, a porcelain figure on an end table next to the couch began to "wiggle." It then flew two feet into the room and landed on the rug with a loud crashing sound, but was unbroken. The cousin observed this from the other end of the room; no one else was there. She said that it "moved so fast that it looked like a small feather."

Another poltergeist case with a similar level of authenticity was investigated by Roll in Miami. It involved the inexplicable movements of small novelty items in a store such as beer mugs, alligator ashtrays and wallets around the room in which they were being stored. Many of them apparently slid off their shelves and fell to the floor, breaking in the process. The store's owners were naturally worried at the amount of damage occurring, and indeed these strange movements caused chaos in the store. As the manager said, "For three days we picked things up off the floor as fast as they would fall down. It was going on all day—quite violently —but not hurting anything. . . . We tried to keep it quiet because we knew it would hurt our business, because we are right in the middle of a season—and it would draw a bunch of curiosity seekers and the like, so we tried to keep it quiet for about four days. Then finally, delivery men saw those things happening and people coming in and out would see it happen and word got out and there were more and more people coming. And somebody suggested that with the girls crying in the shop from fright we had better notify the police, so I did."

When the police arrived they witnessed a number of objects falling off the shelves when all the staff were some distance away. A local professional magician was also called in, and he could detect no "invisible strings" or other tricks that could have been used. Nor could the objects have slid off on their own due to vibrations or movements of the shelves; all the shelves but one were shaken purposely but nothing fell from them.

A number of investigations were made, especially by Roll and his colleague, J. G. Pratt. They catalogued a total of 224 such abnormal movements, as recorded by various people, 52 incidents having occurred on a single day alone. In one case two people saw a bottle move as they watched it from different directions, and described it independently. As one of them said, "It went way out in the air, then down with a bang on its neck. Then it bounced on its side three times." The other witness noted that it moved similarly and that the other clerks were some distance away from the bottle.

One of the most interesting things about the Miami disturbance was that it proved possible to make tests of how the particular objects moved in certain cases. Thus Roll wrote: "I wanted to find out if the objects moved directly off the shelves or if they could be made to rise up in the air. I had therefore placed some other objects, including some notebooks, in front of a large glass. These were undisturbed, so the glass must have moved up in the air at least two inches to have cleared them." Various other tests of a similar kind were performed, each of which involved inexplicable movements of the objects. In particular, it was found that objects were most often disturbed in particular places. In several instances items placed at such locations and kept under careful surveillance were disturbed. The movements only seemed to occur when one of the clerks, a young man of nineteen, was present. Roll and Pratt concluded: "In no instance was evidence found indicating that any of the events were caused fraudulently or in any recognized nonpersonal way."

None of these cases, however, have been authenticated in as much convincing detail as the one described in Chapter 2 concerning the Rosenheim affair.

Whatever psychic powers the young girl at Rosenheim may have possessed, she had at least one thing in common with Nina Kulagina, a housewife born in Leningrad in the mid-1920s, whose ability to make objects move has been investigated extensively by Russian scientists and also by four parapsychologists from the West—H. Keil, B. Herbert, J. G. Pratt and M. Ullman.

This trait is reported by the Western investigators as follows:

Kulagina can, by placing her hands on a person's forearm, induce a sensation that feels like very real heat to the point of being painful. There are differences between individuals as to how severe the heat appears to be, and one person may experience substantial variations from trial to trial. . . . Herbert reported unbearable pain which he continued to endure only as a personal sacrifice for science, so to speak. Cassirer felt nothing on one occasion and real heat sensation on another. Keil and Fahler felt strong heat and pain sensations but not to a degree that made it difficult to endure them. Both Herbert and Fahler had "burn" marks on their arms which were visible for several hours. No blisters or other negative aftereffects developed.

Further tests with a mercury thermometer showed that there was no increase of skin temperature even though this sensation of heat was experienced by the investigator. A piece of lead partly shielding Kulagina's hand as it was placed on the investigator's arm also reduced the heating sensation.

Direct fraud can be excluded here quite effectively. A burn mark might have been produced on the bare skin by application of a suitably palmed irritant, but one of the investigators was affected by Kulagina without direct contact occurring; there was always a gap of at least two centimeters. In any case Kulagina's hands had been carefully examined beforehand and were found to be completely dry.

Films have also been taken of her causing the movement of various objects by repeated hand movements or even, it is claimed in some cases, very little (if any) motion of Kulagina's body or arms.

To quote again from the report of the Western observers: "Kulagina found it relatively easy to move long objects in an upright position, such as cigar containers made from a thin non-magnetic material, tall glass objects and even cigarettes standing on end . . . continuous sliding movements usually last for a fraction of a second to a few seconds. While Kulagina has moved objects over longer distances of the order of thirty to forty centimeters, this seldom happened as one continuous movement. . . . The size of objects varies from a single match to large glasses and other containers including a ten-centimeter Plexiglass cube." The surface on which these movements occurred ". . . varied from glass and Plexiglass to wooden tabletops sometimes covered with a tablecloth"; there seemed little difference in the movements achieved on these different surfaces. Kulagina has also affected the motion of objects suspended in a plastic cube, such as a Ping-Pong ball at the end of a light suspension spring.

The ability of another Russian woman, Alla Vinogradova, to move objects has been described in Chapter 2. She could also move a range of other objects, such as plastic pen caps, Ping-Pong balls, various coins and wooden matches. She caused them to move on the same surface of the large transparent plastic cube which served as the table in the earlier investigation of the rolling cigar tube. Thus: "She [Alla] could roll the pen caps in bumpy fashion due to their pocket clip or could make them skip without rolling. She could push them over from a standing position but could not raise them upright from a prone position. The Ping-Pong ball could be made to skid across the table or to roll or bounce."

These effects have been filmed on numerous occasions, with great care taken to exclude fraud. They are repeatable at a very high level, so much so that they can even be duplicated by certain others. Dr. Vinogradova, Alla's husband, has instigated a training program for developing this psychokinetic power in suitably chosen subjects. And the phenomena indeed seem to be paranormal. How can science explain the way the matches moved under Alla's power: "The matches, which were emptied out of their box in a jumbled heap on the table, reacted in even stranger fashion. At first the jumbled mass did not move as a whole. As Alla pushed at them from about a foot away, individual matches suddenly jumped, scattering the other matches around them. A momentum appeared to be building up that reminded me of popcorn held over a fire. The wooden matches 'popped' in all directions." Here surely is a paranormal phenomenon, one outside the bounds of modern physics.

A nineteenth-century case that aroused an enormous amount of interest throughout the United States and in Europe, and still of relevance today on account of the amount of investigation and documentation, concerned two sisters, Margaretta and Catherine Fox.

It started on December 11, 1847, in a tiny wooden house in the village of Hydesville, New York. The house was occupied by its new tenant, a farmer named John Fox, his wife and their two daughters, Margaretta aged fourteen and Catherine aged twelve. The previous tenant of the house had left it because of mysterious noises but these did not become noticeable to the new occupants till March 1848. Raps, knocks and noises as of moving furniture were heard at night and increased so much in intensity as to seriously disturb the family's sleep. The mother signed a deposition on April 11 of that year describing the nature of the occurrences.

On the night of the first disturbance we all got up, lighted a candle and searched the entire house, the noises continuing during the time and being heard near the same place. Although not very loud, it produced a jar of the bedsteads and chairs that could be felt when we were in bed. It was a tremulous motion more than a sudden jar. We could feel the jar when standing on the floor. It continued on this night until we slept. I did not sleep until about twelve o'clock. On March 30 we were disturbed all night. The noises were heard in all parts of the house. My husband stationed himself outside the door while I stood inside and the knocks came on the door between us. We heard footsteps in the pantry and walking downstairs; we could not rest, and I then concluded that the house must be haunted by some unhappy restless spirit. I had often heard of such things but had never witnessed anything of that kind that I could not account for before. On Friday night, March 31, 1848, we concluded to go to bed early and not permit ourselves to be disturbed by the noises, but try and get a night's rest. My husband was here on all these occasions and helped in the search. It was very early when we went to bed on this night—hardly dark. I had been so broken of my rest that I was almost sick. My husband had not gone to bed when we first heard the noise on this evening. I had just lain down. It commenced as usual.

But that evening the family found that, by a simple code, they could apparently get an intelligent response to questions posed to the "rapper," even to the extent of finding that the cause of the knocks was apparently a peddler who had been murdered in the house some years previously and his body buried in the cellar. Neighbors then called in also had their questions answered by the "rapper." Attempts were made to dig up the cellar floor, where some human remains were discovered; some fifty years later part of the wall of the cellar collapsed and an almost complete human skeleton and a peddler's tin box were then claimed to have been discovered. This is controversial, however, there being no proof that the bones were of human origin, and it was even rumored that they had been planted by the owner in an effort to arouse interest in the house, which by then had become a spiritualist's shrine.

A third and older sister of Margaretta and Catherine, called Leah, also became involved when the mother and her two daughters moved to the older sister's house. The raps and knocks went with them and were soon apparently also caused by Leah herself. Furthermore, there burst out such a rash of noises in the houses of people who had heard of the case that within a few years many thousands were regarded as able to produce the raps themselves. Above all, the noises were produced countless times and under a wide variety of conditions by the three Fox sisters over a period extending almost to their deaths in the early 1890s.

Surely here was a phenomenon recurring often enough to be able to produce an explanation. Since table rapping soon blossomed forth to include a whole array of paranormal phenomena —table moving, materialization, messages from the dead almost on demand and numerous other phenomena—an explanation of the noises could also be an important first step to explaining these other strange events, if they are not fraudulent.

What of the companion psychic phenomenon—table moving? This has almost as long a history as noises and even more careful investigation. It is to be expected that here is a more clear-cut test, since it would seem easy to detect if a table is being moved by physical force or not. If there is no one touching the table as it moves, then either carefully prepared cheating is being employed (which should be possible to discover by careful analysis) or a truly paranormal phenomenon has occurred.

There are reported cases of that sort, such as the one involving two devoted spiritualists, the Reverend Stainton Moses and Sergeant W. Cox. It took place in 1873 at Cox's house, starting with frequent loud rappings on the table as Moses read the *Times* and Cox opened letters. Table movement occurred when Moses put one arm on the table, it swaying with an "ague fit" and moving forward a few inches. The two then held their hands about eight inches above the table, when it supposedly rose three inches on one side and three inches on the other.

Many cases have been reported, even up to the present day, of a table moving apparently inexplicably in such a situation. The group of psychic investigators in Toronto who fantasized about an imaginary English knight called Philip living in the reign of Charles I reported recently one of their experiences of this sort: "However, after a few moments, the raps started, and then the table began to stand on one leg and to move about the room. (Sometimes this was a straight movement on one leg; at other times the table would revolve in a sort of 'waltzing' movement.) The table also tended to turn on its side and slide along on just two legs, with the other two in the air. During all these gyrations the group members, as far as possible, kept their hands on the top of the table, in whatever position that 'top' might be. But even so, the table, on two legs, would move away from them or toward them in a completely random fashion."

The various movements of the table were not always easy to explain as being caused by physical force applied by members of the group. Thus, at one time: "Although all the sitters were standing up, their hands on top of the table, the table turned over gently and then just flipped upside down with all its legs in the air." On that evening "some members of the group were convinced that all four legs of the table were off the ground at the same time." But a later attempt to capture such a levitation on film failed. As the report runs: "At the end of the evening the table was levitated fully about half an inch from the floor. It floated a short distance, in the presence of two independent witnesses. Although there was a cameraman present it was almost impossible to get a shot of this because of the angle."

The "Philip" group has continued its activity with table-moving and table-rapping powers apparently undiminished. Indeed "Philip" appeared for a local television station in Canada and performed quite remarkably. As one of the group wrote, "Philip quickly made it obvious that he felt his place was on the dais with the speakers' panel, rather than in the corner of the studio, and his efforts to mount the dais and take his 'rightful' place were hilarious to say the least." The table apparently wriggled and pushed to get to the front of the dais, finally succeeding. Having done so, the program moderator said "Hallo" to "Philip," whereupon to his surprise a very loud rap in response came from the table, right underneath his hand. The moderator even carried on a "conversation" with the table for several minutes, the table responding with loud and clear raps all that time. The conclusion of these activities, according to a Philip group member, is still uncertain. "Although on occasion the table has appeared to be floating, it has never lifted sufficiently for the group to be sure of this." The noises or raps continue loud and clear—these being the most easily produced phenomena.

Others have turned to processes similar to table rapping or moving in an attempt to discover how extensive is the range of human powers in moving objects. There has been special interest in the pioneering work of J. B. Rhine at his Institute for Parapsychology at Duke University in North Carolina. Rhine and his colleagues tried to detect any control that people might exert over random events such as the toss of a coin or the throw of a die. Work on this was started by Rhine in 1934 and has continued at his institute and elsewhere since then. The outcome of this was the claim that "not just a few gamblers but a large fraction of the population can mentally influence the fall of dice," as was remarked recently by Dr. Helmut Schmidt, the director of Rhine's Institute for Parapsychology for several years.

These die-throwing tests were conducted carefully. Hand throwing of the die was soon replaced by the use of a cup and then an automatic mechanical die tumbler which stopped automatically at regular intervals, at each of which the uppermost face could be noted. The earlier positive results were continued with the automatic die thrower, with results which were suggestive but not compelling. For example, Rhine reported in 1943 on nine hundred trials which were made with two dice in the automatic tumbler over a two-day period. The three adults "willing" a particular face of the die to appear obtained success in 19.2 percent of their attempts, whereas pure chance would give a 16.7 percent success rate. It can be shown that such a success rate would be expected to occur once in four hundred such tests by pure chance, which indicates that chance is a rather unlikely explanation but not completely impossible. The statistical analysis of such results has shown a "decline" effect—after the first twenty or thirty attempts a subject will attain only chance level. This has been claimed by Rhine to indicate that a true paranormal effect is occurring, the decline of success being caused by psychological factors such as boredom and fatigue. Thus Rhine wrote in 1948 in his book *The Reach of the Mind:* ". . . it was found that in the high dice tests nearly all the extra-chance hits were in the first run of the two- and three-run sets" (a run being of twenty-four single die readings). Dice bias was ruled out as an explanation of this decline effect, as were errors by the experimenters. Rhine concluded, "Happily the experiment was better than we knew and, with the confirmatory results from the score decline found later, it fully meets the requirements for the evidence of PK effects."

The evidence presented earlier for table moving was not very strong, since the amount of pressure applied by the subjects to the table was not monitored in the cases described there.

Thus proper evidence is unavailable from the "Philip" group's efforts and from those of others with like interest. Such objective evidence of pressure recording is essential in this phenomenon after the investigation of table moving by the eminent British scientist Michael Faraday in the last century. He put pellets of wax mixed with turpentine on the underside of a piece of cardboard which he then placed on a table, and he observed "by the displacement of the pellets that the hands had moved further than the table and that the latter had dragged behind; that the hands, in fact, had pushed the card to the left, and that the table had followed and been dragged by it." He concluded that table moving was achieved by unconscious muscular action.

His conclusion is as valid now as it was then. This is not to say that *all* table moving is caused in such a simple, though subtle, manner. But clearly the onus is on the investigator to prove that such unconscious muscular activity is not at work. Pellets such as Faraday used, or more recently developed equipment such as electromyographs attached to the sitters' arms to measure the electrical by-product of muscular activity, are a sine qua non of such tests involving direct contact by sitters' hands on the table. There has already been preliminary work on this by groups in England, especially those led by K. J. Batcheldor and later by Colin Brookes-Smith (from which the "Philip" group in Toronto took many hints). Measurements of mechanical force applied to the table were taken by strain gauges during the tests, though they have not given any validation of the phenomenon.

The die-throwing tests of Rhine and others, described in this chapter, were claimed as authentications of the power of mind over matter by their authors. Yet one of the experts in this field, Dr. Helmut Schmidt, has written: "For a quantitative study of PK, dice may not be the ideal tool, since the physical and mathematical analysis of rolling dice, even in the absence of PK, presents a mathematically very complex problem." Thus die throwing is another area which is not completely satisfactory. Only if phenomenally high success rates had been achieved under very careful conditions would the control of die throwing be eligible for authentication of PK; that has not occurred.

There is another reason why the results obtained from die throwing must be treated with caution. The decline effect was cited as support for psychic control of the die. But this is not really so, since the decline effect may also be produced by taking too small a sample and expecting chance results to be valid there. This question does not seem to have been investigated in enough detail and so leaves the problem open. This decline effect is in any case suspicious since the successes in some automated tests are known to happen whether or not the subject is "willing" anything to occur, or is even present or thinking about the test at all. This must indicate that at least part of the apparent success lies in a failure to apply the laws of chance properly.

It could also be possible that electric fields are involved in die throwing. Thus, if one part of the surface of a die had more electric charge than the remainder, it could be affected by its surroundings in an asymmetrical manner. The decline effect could be explained on these grounds, since it would take time for the charge to leak off the die. The charges and fields necessary to produce appreciable effects in die throwing may not be at all as high as that required for table turning or poltergeist activity. This possibility becomes more realistic when we turn to the powers of object moving possessed by Alla Vinogradova and Kulagina.

We accepted that there was unlikely to be fraud in these cases, especially since the investigator, Pamela de Maigret, discovered she could achieve the same effect when visiting Vinogradova. There are certain features which indicate, however, that Alla probably uses the force of electrostatics. This force is that arising when electric charges are produced by friction during removal of nylon clothing or combing the hair with a plastic comb. Small pieces of hair can be seen to be "attracted" to the comb after such activity, the force being that of electrostatics between the electric charge produced on the comb by friction with the hair and that on the hairs themselves. Alla does indeed use a certain amount of rubbing, both of her hands, which she then places near the object to be moved, and of the plastic tabletop. This latter would be a particularly good repository for electric charge, which could then be transferred to the various objects. These would then be repulsed by her charged hands.

It is very likely that electrostatics is all that is needed to explain Alla Vinogradova's apparently paranormal feats. An electrostatic component is accepted by her husband Adamenko as being important. A group of four Western parapsychologists who have paid a number of visits to Russia to investigate this and other similar cases reported recently: "At this stage, however, it is somewhat difficult to separate the clearly paranormal aspects of the movement of objects by Vinogradova from those which may be due to the normal although highly skilled usage of static electricity."

A further feature of Vinogradova's powers which seem to indicate normalcy rather than otherwise is the use of a plastic tabletop. Plastic (and glass) are remarkable materials for holding electric charge on their surface. I have myself, in collaboration with E. Balanovski, discovered this after a certain amount of careful study of what, on the face of it, appeared to be a perfect example of a paranormal phenomenon. It involved the rotation of a needle or other thin strip of material such as wood, metal or plastic when it was suspended by a thin thread from inside the top of a sealed cylinder of clear plastic. A subject would pass his hands near the cylinder; the strip of material would rotate, sometimes even by as much as 360 degrees—a complete rotation.

Various subjects were discovered to have the ability to cause such rotation. Fraud was ruled out by various methods. Thus a sensitive magnetometer positioned next to the cylinder could detect if a palmed magnet was being used to rotate a magnetizable metal object. This caught out the magician, James Randi (known professionally as "the Amazing Randi") when he claimed to be able to achieve this by means of ordinary stage magic—he had a magnet palmed in his hand, as the magnetometer showed by the large magnetic fields it measured when he moved his hand near the cylinder. Static electrical effects produced by rubbing the hands were also excluded by observing if there were significant changes of electric field near the cylinder as a subject brought his hands nearby. The cylinder was screwed down to a very sturdy bench so that no mechanical movements could produce the effect. Many hours of videotape were taken of the effect —rotation of metal, plastic and wood strips all being observed— and validation was not at all in doubt.

These tests were spread out over several weeks since the subjects were not always readily available. During that period I thought that I had finally obtained a repeatable paranormal phenomenon, one that could occur under the very best laboratory conditions possible and with the high level of validation required by my scientific colleagues. It was indeed a very exciting time!

It was when both I and Balanovski discovered we had the "fluence" and could achieve the rotation, at least on certain days, that doubt as to the true paranormalcy became very strong in my mind. Different types of plastic were used as the outer cylinder and we noticed that they resulted in differing amounts of rotation. Finally, to clinch it, we found that rubbing the outer surface of the cylinder produced the best effect. We discovered that the amount of rotation of a suspended strip of material was proportional to the surface electric charge on the cylinder near the ends of the suspended object. This charge decayed in time, explaining why we (and anyone else) could not achieve the effect without producing charge on the cylinder surface by gently rubbing it, even if only for an instant, after the cylinder had been left for several hours. It also explained the decline effect—why the rotation became progressively worse if the cylinder surface were not touched at all during the tests. The use of this surface charge also explained why subjects could cause rotation even when they were grounded by a wire attached at one end to the earth and at the other to a metal electrode on their skin (usually the ankle).

In our case of object rotation in a plastic cylinder, a crucial test for the electrostatic explanation was that of covering the cylinder with antistatic ointment (leaving the transparency of the cylinder unchanged) so that no static charge could be stored there. The effect was then destroyed completely. No one, not even our best subjects rubbing the cylinder in any way they chose, could produce a glimmer of life in the suspended strip of material. This explanation of object movement seems to apply to Alla Vinogradova's feats. Only if she could move objects on a metal table would this explanation be invalidated. This test has been suggested but, as far as I know, not carried out. Until the result of such a test is known, no firm conclusion can be drawn, though the various features of the case suggest that the phenomenon will disappear.

To summarize, affecting the movement of objects suspended sensitively inside a plastic container can well be explained by electrostatics, unless precautions are taken such as the use of antistatic ointment or careful measurements are made of surface electric charge. The same normal explanation may well be true of the object movement caused by Kulagina, though this is difficult to assess until tests are carried out with her similar to those suggested for Vinogradova, the use of a metal tabletop being critical here. Various electric- and magnetic-field measurements have been made while Kulagina is in action, but they do not seem to rule out the electrostatic explanation.

Finally, my colleague and I looked more carefully into a somewhat different test successfully demonstrated by several English children on apparatus constructed by the English physicist, Professor John Hasted. He formed the shape of a "T" by gluing together two plastic straws and set it so that the "T" was vertical on a circular polystyrene disk floating on water in a small beaker. This was then covered by a large glass jar to prevent currents of air from causing the straws on the disk to rotate. Both a fourteen-year-old boy, Willie G., and a thirteen-year-old girl, Julie K., were able to cause the straws to rotate merely by concentrating on them. As Professor Hasted reported on one test, "After five minutes the plastic straws began to move very slowly in a clockwise direction for 40 degrees." This was another phenomenon which deserved further study. It could not have been caused by fraud, since the straws were not accessible to the subjects. Neither direct mechanical movement of the beaker nor surreptitious "blowing" by the subjects would have been effective since it was covered by the glass jar. Nor could one expect much effect of heat on the straws. "Invisible" strings would also have been very difficult to use without detection.

We arrived at an explanation of this similar to the explanation we would expect for the effects that Kulagina achieves. We discovered that the straws were so sensitive that they would respond to a very light draught of air. The purpose of the glass cover over the beaker was to prevent that, but it could not do so completely. This was due to the simple reason that a source of heat on one side of the jar could produce small convection currents inside the jar; these could easily cause the straw to rotate.

One test was carried out at the home of a thirteen-year-old girl. She was successful in causing the "T" to rotate slowly as she sat and gazed at it. At no time did she touch the outer surface of the glass container, and our measurements showed that no electro-static or radio-wave energy was being used. Yet the straw still rotated, slowly but surely.

The cause became clear when we discovered drops of moisture condensing on one side of the jar. This was caused by heat, coming from an electric fire on the other side of the jar (it was a cold day, so that the fire was needed). The source of heat was enough to cause water in the beaker to evaporate and move away to the side of the jar farthest away from the fire, and there condense. We estimated that this would have produced a convection current more than enough to cause the observed rotation of the "T." Confirmation of this mechanism was forthcoming when the test was performed in an equably heated room at Bath University with no strong temperature gradients: no rotation occurred at all.

We can now summarize the position reached so far. It is impossible to explain poltergeists, table moving and table rapping by any electromagnetic mechanism, since too much energy would have to be transferred to be consistent with a lack of side effects of such energy transfer. In any event the latter two of these phenomena do not have very strong evidence to support them; nor does the influencing of die throwing, which in any case has a possible electrostatic explanation, as does the movement of small objects. This latter phenomenon, in the form of rotation, was tested under laboratory conditions and indeed found to be caused either by electrostatics or by heat effects (according to the apparatus being used).

Other reported cases of small object movement, as well as some investigated by the author, were ascribed to fraud. Thus we are left with poltergeists as presently inexplicable according to modern science. How can we face up to that?

One crucial feature of the Rosenheim case may be in this "inexplicable" category. The electric current meter used during the disturbances recorded strong deflections which, however, were mechanically made. The record was "in the form of a loop" and looked as if it had been made mechanically; the recording paper was even torn, the recording needle having been pressed down with some force. Since the deflections were not actually observed in the process of being recorded, it is not possible to say how they were made. But a human hand seems most likely.

Some cases may be the result of deliberate mischief. Various incidents investigated during the "poltergeist season" in the 1890s confirmed this. One in particular, investigated by a General Pitt-Rivers, turned out to be a joint effort of a number of children. They confessed quickly enough when the General invited the police to take action, saying that he and his wife would give evidence of fraud. Children are more lively and inventive than is normally imagined, and as Ronald Pearsall so aptly summed up, ". . . even more intelligent than the classic dupes—vicars and policemen—who stumbled through rustic poltergeist phenomena. . . . and it must have been highly enjoyable for children to throw stones and break windows, tear up snowdrops, toss cats into fires, knowing that they would not be punished."

The Rosenheim case cannot have been quite so simple. Yet even here we have to depend mainly on the reports of eyewitnesses. As many legal cases have made clear, even highly trained people become poorer observers under stress; the advent of poltergeist-like phenomena would undoubtedly be stressful to many. A typical case of poor reliability presented by the psychologist Robert Buckhout was of a police officer who testified that he saw a defendant, a black man, shoot a victim as both stood in a doorway 120 feet away. But Buckhout added, "Checking for the defense we found the scene so poorly lit that we could hardly see a person's silhouette let alone the face."

Even the level of accuracy of reporting in psychic cases is poor. This is somewhat to be expected in newspaper accounts of them. Yet exception has to be taken of the account in which it was stated categorically *and in italics* that the Dutch clairvoyant Gerard Croiset had been able to pinpoint the spot where a murdered girl's body was to be found and that "the body was later discovered there. . . ." No such thing happened in this case (the body never being discovered at all), as I learned on discussing the matter with one of the people closely involved with it.

The only possible explanation left open to us in this whole poltergeist phenomenon is that of a mixture of expectation, hallucination and trickery. This has often occurred in such cases, where the central character around whom the events revolve tends to be of a highly neurotic nature. It may even be that such a subject is performing the "supernatural" feats in a natural way to draw attention to his or her distress.

Such an explanation is the only one which seems to fit in with a scientific view of the world. It is also consistent with other poltergeist cases. Yet it cannot be checked because of the elusive nature of the poltergeist effect. Until that changes, the only possibility left open to us is the one stated above, involving expectation, hallucination and trickery. In addition, along with inaccurate reporting, many seemingly paranormal phenomena also contain such disparate ingredients as coincidence, fantasy, suppressed memory and the need to believe. These ingredients will be considered in the chapters that follow.

9

The Framework of the Paranormal: Fraud and Mischief

THE PROBLEM OF FRAUD IN PSYCHIC PHENOMENA HAS A LONG and painful history. In the middle and latter part of the last century a great deal of trickery was uncovered, the frauds being perpetrated under cover of darkness—usually required by mediums in the performance of their task. A typical exposure occurred at a séance given by a Mrs. Mellon on October 12, 1894, when a Mr. Henry suddenly seized Mrs. Mellon in the cabinet in which she had secreted herself, just as what looked like a spirit was emerging from it. Mrs. Mellon was discovered on her knees, with muslinlike material around her head and shoulders. She struggled while Mr. Henry asked for more light; when this appeared Mrs. Mellon could be seen to have a black mask over her face. At that moment the light was blown out and Mr. Henry was attacked by Mrs. Mellon's supporters, led by her husband. A false beard, a small black shawl, some old muslin stockings and other paraphernalia were later discovered in the cabinet.

One "medium," on finally giving up in disgust, is quoted as saying early in 1977, "For thirteen revolting but highly profitable years I was a stellar member of a network of about two thousand phony mediums spread all over the country. We traded information about clients, conspiring to cheat countless people out of millions of dollars." He added, "Every fake medium maintains personal information files on each client. Before I'd meet a client from Chicago I'd call mediums in that city until I found one with information on that particular client." He then went on to describe how, dressed all in black, he would walk around the darkened séance room causing trumpets and other objects to float around, and even to levitate his clients. "My partner and I—dressed in black so we'd be invisible in the darkened room—simply grasped the legs of the client's chair and lifted him into the air. Not once did anyone catch on, although they could have."

The problem of fraud is undoubtedly still present. A recent example involved six children aged from eight to thirteen, each of whom had claimed to be able to bend spoons like Uri Geller. They were tested by a group from the University of Bath in a room with a one-way viewing mirror so that they could be watched without their knowledge. There were also observers in the room who pretended they were running the tests. As the report on the tests indicates: "The observers in the room were instructed to deliberately relax their vigilance at intervals after the first twenty minutes. The experimenters were specially alert during these periods and in all cases except Subject C, a thirteen-year-old girl, they observed and photographed cheating by the subjects. Subject A put the rod under her foot to bend it; B, E and F used two hands to bend the spoon using considerable muscular power, while D tried to hide his hands under a table to bend a spoon in both hands out of sight of the observer." We note the similarity of ages between these children and the two younger Fox sisters.

The children were definitely cheating because they were under pressure to achieve the "Geller effect" (spoon bending) to order. Little wonder that one serious psychic investigator has said, "The history of materialization, so far as professional mediums are concerned, is practically one unbroken line of fraud." Nor has fraud been absent from all other types of psychic phenomena— table turning or rapping, clairvoyance, levitation, voices from the dead, telepathy, spirit writing and so on. Even the subjects who are regarded as having initiated recent Western interest in the supernatural, the Fox sisters described in the previous chapter, are not above suspicion. The Fox sisters' fame spread like wildfire, and they gave demonstrations to large audiences, first in America and then later in Europe.

But in 1888 the two younger sisters made public confessions explaining how the raps were produced on stage. However, the following year the elder of the two sisters retracted her confession; both sisters died shortly afterward. The confessions may have been made out of spite toward their older sister, Leah, with whom there had been problems, though the story has never been completely clarified.

Fraud is not restricted to the performers. Even experimenters are not above suspicion. One of them was caught red-handed recently by his colleagues, who had become suspicious of some of the results of an experiment he was running. When they kept a careful watch they saw him changing the measurements; the particular experimenter was forced to resign. The case was even more distressful since the particular person involved was director of one of the most famous research institutes of parapsychology. Yet he had given in to pressure from outside to produce results proving telepathy.

A typical case of fraud concerning UFOs was told me by one of the American government-sponsored UFO team working under E. Condon. A man had sent him a quite remarkable photograph of a "flying saucer." Careful investigation of the locale in which the photo was taken and of the photograph itself (namely the level of focus of various details) led him to conclude that the photograph was of a real saucer which had been thrown into the air and photographed while in flight. Challenged with this interpretation, the photographer finally admitted that was indeed the case. He had asked his son to throw the saucer into the air while he photographed it. He had hoped that the fame he would gain for the "sighting" would persuade his estranged wife to return to him.

In November 1973 an epidemic of spoon bending broke out throughout England, hundreds of reports circulating about the distortion or fracture not only of spoons, forks and knives but also of other household articles such as pans, soup ladles and metal combs. One girl of eleven reported placing a spoon on the sitting-room floor and "concentrating" on it without even touching it. With her mother sitting on the sofa nearby the girl literally "willed" the spoon to bend over until it did so.

In another case, the managing director of a large firm was surprised to discover that the door key in his pocket, on which he had briefly concentrated, had bent so far that he could no longer use it to open his front door. At no time, he recounted, had he applied any force to the key. At a later time he sat with his family in the kitchen of his house, again "willing" the spoon he held in his hand to bend. After an hour without success the family then discovered that a number of articles in various parts of the kitchen had indeed become distorted, including the aerial of a radio, two forks in a cutlery drawer and various other metal household effects.

One lady complained bitterly about the result in her own home, for the piece of gold jewelry specially fashioned for her daughter's eighteenth birthday had been so distorted, she claimed, that it was no longer possible to present it to the girl. Nor was a similar one available before the presentation ceremony on the following day. The mother and family were very distressed at the destruction of the gift and the disappointment caused thereby.

The phenomenon of spoon bending spread quickly throughout Europe, the cases reported being very similar to those in England but sometimes going far beyond them in the extent of destruction which occurred. An elderly German lady recounted how she heard "scratching and scrabbling" in one of the drawers in her cupboard. She went and opened it only to find to her consternation that all the cutlery in it was wriggling around "like snakes." This continued as she watched in fascination. To prove to herself that she was not dreaming or hallucinating in any way she called in two passing police officers who also, so it was reported, saw the cutlery waving backward and forward until many pieces had broken. There seemed to be no question of hallucination, though it would certainly have been easier to explain if such had been the case.

This initial rash of spoon bending was very similar to the epidemic of table and wall rapping. In that case it was sparked off by the Fox sisters described previously; in the case of spoon bending it was caused by the young man called Uri Geller. As with the Fox sisters his name soon became very well known indeed, much more so because the existence of radio and television allowed his supposed powers of spoon bending to be exposed to anyone who happened to have their sets switched on.

Uri Geller claimed there were no conjuring tricks involved in his "act," so that the effect of his demonstration was much more powerful. Indeed, it was this feat, more than the picture guessing or watch starting, that seemed to me so impressive and in conflict with what I knew to be scientifically viable. In addition, there were the many people who came forward to claim that they also could bend spoons.

The subjects who came forward in the various countries which Geller had visited—nearly every one of those in the Western world as well as on the American continents, both north and south—appeared to have no special feature to single them out from the rest of the populace. A good proportion of them proved to be young, below the age of twenty and even down to the age of only three years. The majority had seen Geller on television or heard him on radio and then tried to bend a spoon as he had urged them. But some of the hundreds who came forward had only discovered their "powers" at the urging of a friend. Nor was there any pattern in their manner of achieving success. Some held the spoon in one hand and rubbed it, even quite vigorously in some cases, with the other hand. Others used only one hand in the whole process. There were even those who disdained direct contact and, so they claimed, caused objects to distort in adjacent rooms of their home. The range of objects so distorted proved as wide as the range of subjects: spoons, forks, knives, pots and pans, rings, radio aerials, watch hands, nails, iron bars and ultimately strips of metal of various sorts—aluminum, brass, copper, steel and even rods of lead and tin. Some of these pieces were too rigid to have been bent by hand, though others were extremely flimsy. Not only metal objects proved vulnerable; plastic and wood were the constituents of other things, such as toothbrushes and hairbrushes, bent or shattered by the attentions of the spoon benders. Nor was there any obvious answer as to the most appropriate shape of the object being attacked.

This variability is underlined by a well-established business-man who wrote to a colleague soon after of his experiences: "At lunchtime last Friday I was listening to the Jimmy Young program on the radio. Although having my lunch at the time, as is now my habit whenever any radio or TV program concerning Geller is on, I gently rubbed a dessert spoon lying on the table. Not only did the spoon bend at right angles but two other kitchen (I was eating in the kitchen) articles bent. A stainless steel chip pan now has a handle much more distorted than I saw at one of your Hilton lunches, in fact it appears as if a giant hand had screwed it up, and another kitchen gadget has been folded in two. The chip pan seems to me to be so terribly damaged that if caused by human means some mark would be visible."

The first point, in order of priority, to clarify about the phenomenon was the presence or absence of deception. If all was fraud, then any further investigation would be a sheer waste of time. Claims were made that Geller himself was a conjurer and entertainer who had spent a number of years in his native country of Israel perfecting his act of picture guessing and watch starting and stopping. The account given by an acknowledged conjurer, "the Amazing Randi," is very apposite here: "Leaving the army after minor injuries he was employed as a fashion model for a time, then was a camp counselor, at which job he met Shimson (Shipi) Shtrang, several years his junior. The two happened upon a book that dealt with magic and magicians and began working together on the subject. An act developed and they began working at the kibbutzim and private parties and in nightclubs, claiming supernatural abilities for what was essentially a two-person 'code' routine. Eventually they were brought to court for using the words 'psychokinesis,' 'ESP' and 'parapsychology' in their promotion, and from then on they were not allowed to use such terminology, since they were performing conjuring tricks. This, coupled with the exposure that Geller had faked a photograph of himself with Sophia Loren for the Israeli newspapers, led to Geller's decline in his own country." Clearly Geller's past is not as white as the driven snow. Fraud may be involved in anything he does. It need not be, but clearly caution must be exercised.

It is not to be expected that the many subjects with spoon-bending powers who have come forward in the last few years will be as experienced as Geller at various conjuring tricks which may convince friends and neighbors that one can bend spoons. This would seem especially true of children, whose powers at deception might be expected to be minimal. Yet this is not so, as was shown most clearly by the work of researchers at the University of Bath in England, reported on earlier in this chapter.

Since that time the one-way viewing system has been used a number of times with other subjects, sometimes detecting blatant fraud, at other times not. One of the difficulties in such tests is that less obvious techniques of applying mechanical force can be used by people who are sufficiently strong to bend the object they are attacking. They may hold the object in both hands and apply just enough force to achieve distortion without the force used being noticeable to others until the damage has been done. It is here that one sees how essential it is to have further means of assessing the pressure being applied; or preventing it being applied altogether, by using very strong metal rods; or even disallowing any contact at all between subject and object.

Having realized the problems of fraud, it is clearly necessary to enumerate the various ways that fraud could be used to achieve metal bending, and how they might be guarded against. A number of conjurers have developed a range of techniques for fraudulent spoon bending and undoubtedly they will do so afresh in the future. This list is therefore not meant to be completely exhaustive, but it should cover the major methods available to those not so mature in conjuring as the members of the Magic Circle in Britain or their colleagues in other countries.

One obvious method is that of substitution. An already bent duplicate of the object to be deformed is secreted on the subject's person. He or she then proceeds to distract the audience by suitable means most appropriate to the situation, such as by directing attention to an object being held by someone present, perhaps by saying that it is bending. During the period that close perusal of the subject ceases, the subject switches the object he or she is trying to bend for that already bent. The bend itself is not then immediately shown, but a little time is allowed to elapse to enable attention to return and the sense of close scrutiny of the subject regained. The bend is then slowly revealed as if it was just occurring.

There is one disadvantage against this method of fraud; the subject will have to have things very much under his own control, otherwise he will be handed objects to bend for which he will not have already-bent duplicates. Indeed household spoons and keys come in such a range that it would be very hard to duplicate all of them without the subject "clanking" like an itinerant iron-monger when he moved around. A better way of cheating is to apply force mechanically to the object when the distraction trick has been used, again with delayed exposure of the bend. I have myself observed this type of fraud perpetrated by various sub-jects who have come forward. It can be guarded against in various ways.

First, only objects which are impossible to bend by hand should be used. Since paranormal metal-bending powers might not go beyond those of normal strength, subjects would not perhaps be able to succeed with such strong objects. Thus, in order to test lower levels of paranormal strength over spoons, either the pressure applied should be monitored throughout the test, or the object should be bent in a manner impossible by normal mechanical means (such as against the pressure), or there should be no contact whatever between subject and object. Metal objects in well-sealed containers provide a case in point.

One or another of these approaches must be used to prevent fraud by the application of unnoticed pressure. There are other ways of cheating which are much less likely, such as using chemi-cals virulent enough to act on microcracks on metal surfaces, or the application of thousands of degrees of heat to cause differen-tial expansion or even melting of a spoon handle. Neither of these methods is likely. A further method, requiring hypnotism of the observers (as seems to occur in the Indian rope trick), is also only a faint possibility. These can all be guarded against by various straightforward methods, such as chemical analysis or covering the metal surface with a plastic coating in the first, using a thermometer or other temperature sensor in the second and using a videotape recorder and other automatic recording instru-ments in the third.

I cannot emphasize enough that there is always the possibility of fraud occurring. One recent subject who had made strong claims as to his ability to move objects and bend spoons was caught using "invisible thread" to move light objects around on a table. I have a videotape of the suspicious movements of another subject who had come forward. He is seen as if drawing up an invisible thread attached to his shirt-sleeves; objects then moved around in a bizarre fashion (though understandable enough after analyzing the suspicious hand movements).

Another point to guard against in such investigations is that inaccuracies are constantly creeping into accounts by others. Unless one investigates these very carefully one does not easily detect them.

That the problem of inaccurate reporting is always present wherever psychic events are involved I have learned to my own cost. For example, I was charged by the conjurer, "the Amazing Randi," of having formed most of my opinions of Geller at Birkbeck College. Specifically, he writes in his attack, *The Magic of Uri Geller:* "Arthur Clarke . . . has described the conditions for the tests at Birkbeck in London—where Taylor formed most of his opinions of Geller—as 'incredibly sloppy' and said that he was not impressed by some of the scientists who got involved with Geller." But at no time did I ever set foot inside the hallowed walls of Birkbeck College when Geller was there. I was not invited, nor did I even know that the tests were being conducted until much later. I have no idea how Randi connected me with the Birkbeck scientists, but he was incorrect!

The evidence for paranormal spoon bending presented up to now is suggestive but certainly not watertight. This is the conclusion I have come to more recently on carefully reconsidering the cases which I had investigated personally and which led me earlier to conclude that the phenomenon was truly authentic. Nor is this modified assessment altered by the accounts of work by others who saw subjects with supposed spoon-bending powers.

The conclusion I have now arrived at is based on various criticisms which could be (and have been) made of earlier tests of the phenomenon, in particular that of inadequate documentation of the amount of force being applied to the spoon by a subject during a test. Both videotape and direct recordings of the pressure being used during any spoon-bending session are necessary for authentication. I set out to collect such documentation several years ago.

One of the methods I used was the "letter balance" method, with the metal strip to be bent being screwed down to the top of a letter balance. The pressure applied was monitored continuously during the test by a video camera focused on the needle of the balance. This camera also gave a close-up of the subject's finger gently stroking the strip and recorded a clock used to prove continuity of the record. Another video camera was used for a wider shot to detect suspicious movements. At the same time a range of radio-wave detectors was used.

This method was inspected by various sceptics, including "the Amazing Randi," who commented afterward that the setup would be satisfactory provided the second video recorder was remote enough from the scene of the crime; this was accordingly ensured. Having set up a suitable test, and enlisted the aid of several colleagues, we invited various subjects to perform.

The first to be put under our scrutiny was Uri Geller himself, who happened to be in London launching a record he had been involved with. He came to the laboratory for one and a half hours. In spite of the very friendly atmosphere he did not succeed at all during that period. Nor has he returned to be tested again under these (or any other) conditions, in spite of several warm invitations to him to do so. One could suppose that his powers desert him in the presence of sceptics, but during that test at no time did I or any of my team express any form of scepticism; I do not think we even *thought* a harsh thought! As far as I am concerned, there endeth the saga of Uri Geller; if he is not prepared to be tested under such conditions his powers cannot be authentic.

Since that time other subjects have been tested using the same protocol. Most of these subjects had been observed to "bend spoons" in a decidedly paranormal fashion at earlier times, though they did not include the two boys who have been tested extensively, though separately, by myself and John Hastead, and mentioned in my book *Superminds,* which contained an earlier account of my spoon-bending analysis. One of these refused even to be observed during a test, but had to be in a separate room on his own. The other did not wish to be observed in a laboratory setting, and proved extremely difficult to work with.

The other subjects who came to the laboratory and attempted to "bend spoons" on the letter balance failed completely. Nor was there any evidence of radio-wave or electrostatic effects that might have been expected during their attempts. This method of authentication was completely unsuccessful.

In case the laboratory type of atmosphere engendered by the above protocol caused the subjects to lose their psychic power, another technique was developed to assess the amount of muscular force the subjects were using. This involved using a machine called an electromyograph (or EMG) during the test. This machine measures the amount of muscular activity in a given place by amplifying the concomitant electrical signals arising on the surface of the skin. It had, at times, been added to the equipment used to measure the pressure being applied. To reduce stress it was then used in place of the balance method of measuring pressure. The EMG has the advantage of allowing much greater flexibility and freedom to the subject.

Tests were done with a range of subjects. The protocol used was to attach EMGs to both arms of the subject and monitor the level of muscular activity by focusing a video camera on the face of the EMG meter some distance away from the subject. Such a technique would thus allow the subject to avoid the feeling of being "peered at" directly by the video camera.

Again there was absolutely no success. One subject, whom I will call "Alpha," did cause bending of the strips of metal during the test, but analysis of the EMG's records showed that he/she had applied considerable amounts of force to the strips during the few seconds they were supposedly bending paranormally. When Alpha bent a strip mechanically at our request, the level of muscular activity was very similar to that which occurred during the "paranormal" bending. Direct video record, taken simultaneously to the EMG record, of the hands of the subject also showed very suspicious movements; the hands appeared to be bending the strips mechanically at the time the EMG readings indicated greatest muscular activity.

We also used a variety of electromagnetic recording devices to discover any energy transfer from the subject to the spoon. The whole range of appropriate frequencies—static fields of infinite wavelength down to one-centimeter wavelength microwave radiation—was monitored. At no time during any of the tests was any abnormal signal recorded. One can calculate the level of signal which should have been recorded to cause the bends mentioned above. This gives a value at least a billion times above the natural emission in the microwave range and about a million times the body fields in the near-static range, being monitored during the tests. It is impossible that such signals could have been missed by our detectors.

Certain of the subjects were allowed to attempt to bend spoons with no EMGs or letter balance, but only with electromagnetic detectors present. In these more relaxed conditions there were some cases of supposedly paranormal metal bending achieved by certain of the subjects. But at no time was there any electromagnetic signal above normal, even when the bending was occurring. Naturally the bending that did occur could have been achieved mechanically. It is useful to remark that the cases which did happen were regarded by the subjects as paranormal, with no mechanical force being applied.

There have been alternative methods of approaching the problem of validation of the effect without observing the bending in action. Various pieces of metal which have been bent or fractured by subjects supposedly in a paranormal fashion have been analyzed to see if the internal deformations so produced were normal or not. The analysis has proceeded from the level of crystal size, in metals made up of aggregates of small crystals such as copper, down to that of distorted lines of atoms, called dislocations, produced by the bending process. Fracture surfaces have been magnified many thousands of times using electron microscopes.

None of the results obtained so far from all this investigation have proved that anything truly paranormal has occurred. Thus, certain supposed abnormalities in fracture surfaces have not been universally accepted as such by fracture experts. Some features discovered in specimens do not occur in other similar specimens. Strange aspects, such as the occurrence of higher proportions of certain constituents near a fracture surface than away from it, may have been caused in the manufacture of the item itself. Only carefully prepared specimens can be used without such a problem vitiating the results, and as far as I know that has not been done.

Another technique has been to measure the stresses set up in a metal object while it is being "worked" on by a subject without direct contact. These stresses may be very small indeed and yet are detectable by modern electronic apparatus. To be able to prove from the recordings that something paranormal had happened to the object, it is necessary to show that the measured stresses could not have been achieved by direct mechanical means. At the same time it is necessary to guard carefully against amplification of static electricity produced by people's movements, not a trivial problem with sensitive amplifiers. This method has been claimed, particularly by John Hastead, to give certain validation, but due to the difficulties mentioned above it cannot be said to have done so.

There is no question in my mind that electromagnetism is not at all involved with spoon bending. With my colleague Eduardo Balanovski, I have spent many, many hours working with subjects, with absolutely no success. Indeed I can only conclude from the complete absence of firm results that there is nothing paranormal at all in spoon bending. This leaves unanswered the question as to why subjects have come forward who are apparently honestly convinced that they can achieve paranormal effects.

The most natural explanation may be along the same lines as that of "levitation," which can be played at a party. This is the miraculous feat of four people, at each of the sides of a chair, lifting a person sitting in that chair off the ground, each using only a single finger placed under the seat of the chair. However, this cannot be done by the quartet without some preparation, for if they try the levitation cold, so to say, it is impossible to achieve. Yet if the four people first press down on the head of the seated person for a few minutes, the levitation proceeds with ease.

The simplest understanding of this phenomenon is that the

process of pressing down causes a resetting of the natural level of muscle tension in the arms. It may also block the level of pain signals which can come from the finger during the lifting process. Such an explanation could also be used in the spoon-bending situation. After several minutes of stroking a spoon, sometimes with considerable force being applied (though not enough to cause bending), the subject might not be able to control the amount of pressure he or she is applying. Thus an involuntary muscular effect will be involved, very similar to that discovered by Michael Faraday in the table-moving situation.

That is not the only explanation, of course. Deliberate fraud may be used, and certainly is in some cases. I have already mentioned cheating occurring in various situations where the subjects thought they would not be caught. I even wrote an article on one subject (whom we earlier called Alpha), whom I had observed deliberately bending a spoon. The subject's technique was as good as that of any professional magician. Alpha first ensured that everybody's attention was directed away from the crucial piece of metal he/she held. The spoon was then bent deliberately, but the bent portion hidden in one hand. Alpha then gave the "observers" enough time to redirect their attention to him/her, so that they were all apparently sure that nothing had happened to the precious object. Alpha then said, "I think something is about to happen," as indeed it did. Very gently the bent portion of the metal was brought out from its hiding place in Alpha's hand by his/her other hand. The observers all agreed this was an authentic bend; I did not. But since this process was not filmed in any manner, there was no chance of my own observance of the trick being proved in a way satisfying to all concerned. Had not the observers seen the spoon bend with their own eyes?

More primitive methods of fraud have been used by people who have sent in bent and broken objects claiming they became so "paranormally" during a Geller media appearance. A report by two Americans, Howard Smubler and Marc Seifer, indicates the sort of thing that can happen. The report reads: "In early December we joined a group of friends who were participating in the *National Enquirer*'s duplication of the Geller experiment.

Everyone began concentrating on bending a key and about eight minutes into the experiment a woman in the group stated that her key had bent. The man next to her (whom we shall call Hank) immediately verified that he had seen it bend about 45 degrees. Since this was a trusted group of people, H.S. initially had no reason to doubt the validity of the experience. About a minute later M.S. saw Hank bend another key on the side of his shoe and then hand it to H.S. Because M.S. had witnessed the deception, the fraud was short-lived and the incident was placed in a humorous content."

M.S. was certainly luckier than I was in bringing deception to light. After the tests on Alpha which I described above, I and my colleague wrote a short report which described the method of fraud and our technique of measuring muscle pressure using electromyographs. This was distributed among a dozen scientists who were planning to work with Alpha and also submitted to a scientific journal. However, a hasty retrieval of that manuscript had to be made after numerous telephone calls from Alpha threatening to sue us for libel if we published it. Advice from a lawyer indicated that indeed the article could be interpreted as libelous. Since a jury could have been taken in by such a "psychic," we did not pursue the matter further. Alpha is presently enjoying a great vogue among parapsychologists and a great deal of money has been spent in testing him/her.

I can best finish this chapter by a description given by Smubler and Seifer of how they think some, if not all, of the key- and spoon-bending cases reported by the public-at-large occurred. "Anyone who has participated in a key-bending experiment knows how strong the desire for results becomes. As you close your eyes, holding the keys, you can imagine the metal becoming soft, your fingers apply more pressure and you become convinced that it is in fact bending. With one application of the 'Hank effect' you are the proud owner of a psychically bent object and reap all the psychological and sociological advantages that it can bring. It is irrelevant whether this act is done consciously or unconsciously; the results are the same."

It is also easy to deceive oneself if one has the deep need to believe. Keys bend far more easily than is usually thought. Only a week ago I closely examined my own set of keys, which before-hand I would have sworn were all absolutely straight. Yet on looking at them carefully I found one—my car-door key which I constantly use—was indeed bent through about 5 degrees. How easy for me to think it had been bent by Uri Geller at his last media appearance. Spoon bending—like beauty—is in the eye of the believing beholder!

10

The Framework of the Paranormal: Cues, Fantasy and Memory

A CARDIFF HYPNOTHERAPIST, ARNALL BLOXHAM, HAS PUT MORE than four hundred people under trance and regressed them to earlier "lives." One of the subjects, Mrs. Evans (a pseudonym), was able to recount six different existences during hypnosis, some of which appear to have good grounds for authenticity. In one case Mrs. Evans became Rebecca, a Jewess murdered in York in 1190. She claimed she had died in the crypt of a church in York, later identified by an eminent historian as St. Mary's, Castlegate. This crypt was not, in fact, discovered until some months later when a workman found it under the chancel of the church.

Mrs. Evans also regressed to the fifteenth century in France as a servant named Alison. Her master, Jacques Coeur, did exist as a real person—a wealthy merchant, financier and adviser to Charles VII of France. As Alison, Mrs. Evans said: "At the end of the passage with the portraits and the pictures he has a room where he keeps his porcelain and jade and he has a beautiful golden apple with jewels in it. He said it was given to him by the Sultan of Turkey." No historians could identify the golden apple until one said that he had discovered, in an obscure list of items confiscated from Jacques Coeur by the Treasury, a "grenade" of gold; "grenade" is the French for pomegranate—very much like an apple.

An even earlier "life" of Mrs. Evans was that of Livonia, the wife of a man called Brutus, who was tutor to a prominent Roman family called Constantius on the outskirts of Roman York (which she correctly called Eboracum) about A.D. 286. Constantius's wife was called Helena and his son Constantine; Constantius was supposedly ruler of Britain. Indeed there was a ruler of Britain called Constantius at about that time, and his wife and son were called Helena and Constantine respectively. The only defect in the evidence was that historical records do not indicate that Constantius was governor of Britain in A.D. 286. However, an authority on Roman Britain said that there was a gap in knowledge on the career of Constantius between A.D. 283 and 290, so Mrs. Evans could have been correct.

There are other cases reported by Bloxham with the same level of historical accuracy. It is also remarkable how clearly the previous "lives" are experienced, as if being relived again from memory. For example, when Mrs. Evans relives the death of Livonia it is very vivid, with terror in her voice. The pain people would have experienced during illnesses in earlier lives are also recaptured with great intensity. There are further examples of a similar nature noted by other hypnotists. Undoubtedly these are prima facie strong cases in support of reincarnation.

Belief in reincarnation, in the existence of disembodied spirits and in the possibility of "possession" by them can have terrible consequences. An example is the "Forty Demons Slaughter" in Barnsley, Yorkshire, in 1975. Michael Taylor was thirty-one years of age with five children, and was of below average intelligence. He began to behave wildly after becoming involved with a religious sect and seeing a form of exorcism (casting out of evil spirits) practiced by a woman lay preacher. He attacked the woman, screaming in a frightening manner. A lengthy exorcism was then performed on him, the exorcists claiming they caused forty demons to leave his body. After the exorcism Taylor went home and killed his wife with "unspeakable brutality": he gouged out her eyes, tore out her tongue, almost ripped off her face so that she choked on her own blood. Taylor was found not guilty of murder because of insanity, but at his trial, counsel said, "We submit that those laymen who have been referred to and those clerics in particular who purported to minister to Michael Taylor on that night should be with him in spirit now in this building and each day he is incarcerated in Broadmoor." The belief of the exorcists was a natural extension of the belief in the existence of souls or spirits, as distinct from bodies, to the belief in independent (and sometimes evil) disembodied spirits. "Real ghosts have human minds, but demons have something else. They are an army and their general is Satan," said one of the exorcists.

Let me now turn to a case that is rightly to be termed "possession," and again highly convincing on the surface. Further analysis by a world-renowned authority on reincarnation made it appear even stronger than at first sight. As such, we should consider it in some detail. It involves an elderly Englishman called Edward Ryall, who since the age of eight years or younger was certain that he had lived a previous life which included viewing Halley's comet in 1682.

As Ryall grew older the memories of this earlier life took shape. They were finally written down as the account of the life of John Fletcher, Edward Ryall's alter ego. Fletcher supposedly had been born in 1645 in Taunton, England, and died violently forty years later near his home village of Weston Zoyland in Somerset. The life in between, which would sometimes sweep away Ryall's present experiences, was a full and lusty one dramatically set out in his book *Second Time Round*. It is not on the latter's success as an historical novel but on its historical accuracy that the proof of reincarnation lies in this case. That accuracy turns out to be remarkably good.

An introduction and appendix to Ryall's book were written by Professor Ian Stevenson, who had already made a careful investigation of other reincarnation cases, apparently validating them. Stevenson indicated some of the problems that one faces when trying to assess a case of this type. As in the examples of regression under hypnosis, one has to be sure that the subject could not have had contact with historical material which he had then consciously forgotten, or so-called cryptomnesia. One must also ensure that the subject is not just perpetrating a hoax. If these two possibilities are thoroughly discounted, then reincarnation seems about the only remaining explanation.

Stevenson, with a great deal of experience in the analysis of interviews and what they indicate about a person's psychological makeup, concluded after meeting Ryall: "Conscious, then, that I may be making a mistake, I shall assert my own conviction that this case is no hoax." I would concur with this, though I can only base my judgment on a single meeting with Ryall. But hoax or not, the crucial question is whether Ryall would have had reasonable access to the right historical material.

In order to ask how difficult it would have been for Ryall to have obtained enough material to make John Fletcher's life historically accurate, Stevenson writes: "In short, it seems to me that if Edward Ryall had supplied his mind with all the furnishing necessary for the diverse and often obscure details found in *Second Time Round* he would have shown an intensity and eccentricity of reading habits that his parents would surely have commented about in such a way that he would just as surely have remembered afterward that they had done so."

Having so assessed the situation, Stevenson then analyzed various detailed historical features mentioned in Ryall's book and gave chapter and verse from a number of references to show they were correct. For example, the name of the vicar of the local church in 1685 given by Ryall was verified by the present incumbent. That there was an aurora borealis on the night before the Battle of Sedgemoor (as claimed by Ryall) was verified from Macaulay's *History of England*. Days of the week as well as dates were given correctly by Ryall for some of the events which led up to the battle. Landowning families who figure importantly in the book were indeed correctly described.

The final conclusion arrived at by Stevenson after a painstaking search through nearly fifty historical references was strongly in favor of reincarnation. "In other words," he wrote, "I think it most probable that he has memories of a real previous life and that he is indeed John Fletcher reborn, as he believes himself to be."

This case seemed to have an enormous wealth of data from an earlier life and Professor Stevenson's careful analysis had led him to regard the matter as no hoax. In other words, this was an authentic case of reincarnation. I was myself greatly interested in this conclusion, but felt that further analysis was necessary. Ryall quotes various dates associated with births, marriages or deaths of Fletcher's nearest and dearest. He cites Fletcher's own marriage in 1674 and burial in 1685, the burial of his father in 1660, mother in 1648 and grandfather in 1649, the baptism of his brother in 1647, of his own son on June 30, 1675, and of the daughter of his friend Jeremy Bragg in 1675. There were therefore eight crucial events in John Fletcher's life, each of which should have been recorded in the local village church of Weston Zoyland. Fletcher was a reasonably important man in the locality and was clearly religious. He would undoubtedly have gone through the usual religious ceremonies associated with the various events mentioned. Indeed, in Ryall's book he is quoted as having done so. Thus, concerning his marriage, ". . . it was white gloves for Parson Thomas Perratt [the vicar of Weston Zoyland church at that time] in the year 1674 when the fairest maid of Somerset became my wife. . . ." One would also expect these events to have been recorded by the vicar as part of his duties. Were they in the parish register?

In the summer of 1976 I went down to the church itself and by kind permission of the incumbent, the Reverend Meredith, investigated all of the available records. There were two registers which were most relevant, the clearer of which covered the years from 1644 to 1681. There were only at most half a dozen entries for births, marriages and deaths for each year during that period, and it was easy to see that none of the expected entries mentioned above had been recorded in the register for that period. Nor could they have been written in error a year or two on either side of the expected time, as a quick check showed. The register before 1644 was not so legible, being written in an older script. Yet even that showed no trace of the name Fletcher; nor did the entries under deaths for 1685 in a further register.

The only conclusion I can draw from this rather simple search was that a man called John Fletcher who had supposedly lived at the time and in the way described by Edward Ryall never actually did so. It can only have been a fantasy in Ryall's mind, developed over the years. This could have been nurtured by various historical records which might have been available to Ryall as a boy or a young man. Thus there is an historical novel about a boy who had a beautiful horse (as did John Fletcher) growing up in Taunton. There was also a booklet published in 1929 in aid of the restoration of the Weston Zoyland village church which contained a facsimile of the church records during the Monmouth rebellion of 1685. These may well have come Ryall's way in the wide reading he gives evidence of having done in his life. They could have allowed details to be added to the fantasy life in possibly a quite unconscious manner. At this stage that seems to me the best explanation that fits all the facts of Edward Ryall/ John Fletcher.

There have been some reincarnation cases, as remarkable as the Bloxham ones, reported recently as part of a series of programs made in 1976 by London's Capital Radio and in which I was involved. The subjects were four members of the radio station's staff who, after routine tests, were found to be particularly suggestible. Each of them was hypnotized and regressed to early childhood and then to before birth. They would find themselves in a strange situation and report who they were and what their surroundings were like. They were then taken to ten years later and asked of their experiences; a further ten years were then added, and so on. All the accounts of a given "life" seemed internally consistent. Death experiences at the end of an earlier life were related most dramatically.

One of the typical accounts of an earlier life is as follows. It involved a young man called Alec, in his early twenties. When the process of hypnotic induction had finished, Alec (who was lying on cushions on the floor of the radio studio) had closed his eyes and become very relaxed. At the hypnotist's commands either or both of Alec's arms were raised up to his shoulder level, and then flopped down again upon request. This and other tests appeared to show that he was hypnotized to quite a deep level. Alec was then asked to go back to an earlier time when he was a boy of eight and recount what he saw. This he did, and was then asked to return even earlier to when he was a month old (having been told that he would still be able to understand what was being said to him). At that he curled up on the ground, put his thumb in his mouth and sucked hard, making cooing noises.

Alec was then told by the hypnotist that the time was long before his birth, and was asked what he saw. To the surprise of all present he told how he had his hands manacled after being caught by bandits. He was held in a dark room near his home town of Valencia in Spain in the eighteenth century. After further remarks he was asked to go forward ten years in time and describe again what he saw. He was then the same Spanish nobleman, living with his sister Florence. "Ten years later" he reported he was suffering from terrible pains in his throat and worried about what would happen to Florence when he died. He was then brought back to the present and then out of hypnosis, when he could remember almost nothing of what he had recounted from this earlier "life."

This case is certainly interesting but not detailed enough to be at all convincing, as we remarked at the time. It was also most interesting that after the regression sessions Alec became considerably less tense and an easier person to be with. The use of such sessions in hypnotherapy may indeed be of great value.

A number of the previous lives were interesting for the details they presented of life at that time. Another subject, John, became an eighteen-year-old son of the Duke of Cambridge in 1720. Again, his account was convincing in general but he got a number of important historical dates quite wrong. Thus he claimed that Charles II was on the throne and that the Tories were in power with Gladstone as their prime minister. Another subject, Elizabeth, became an actress living near Stratford in the middle of the seventeenth century. Once again, many historically accurate details were presented, but there were some errors and all the other features could have been read by the subject at some time in the past. Indeed, that is one of the crucial difficulties of assessing any such previous life, not only in the Capital Radio series but also in many other attempts which are currently being made. That information about the historical period has been readily available may itself be incorrectly assessed, as we will see on turning to another approach to reincarnation. Nor do we know that there may not be crucial errors in other "early lives."

Before we turn from analysis of previous "lives" revealed under hypnosis, I would like to add that an attempt was made to discover the depth of hypnosis and the existence of any abnormal brain waves during the "reliving" of these earlier lives. This was done by myself and E. Balanovski in collaboration with Dr. Max Cade and his wife. Skin resistance measurements were used to ascertain the subject's degree of relaxation; the latter was found not to be very great, and indeed the subjects did not appear to be more than lightly hypnotized during the sessions in which I was personally involved. Nor was anything abnormal found on analysis of scalp electrical activity by use of an EEG machine. It could have been that the subjects were somewhat distressed by being hooked up to instruments. They certainly found it very difficult to go back to earlier lives in the experimental setup. They may, on the other hand, have been concerned at being found out in "role playing."

It is appropriate to remark that the evidence presented in the Bloxham tapes is certainly at the same level of dubiety as the Capital Radio results we have just analyzed. There has been a great deal of strong criticism of the former evidence, both of the supposedly remarkable "knowledge" presented and of the naivety of the group involved in making the tapes. The room in which the regressions were carried out has many paintings of historical scenes around its walls, so it is little wonder that subjects could easily regress. Those directly involved accepted all the stories as direct evidence of earlier lives.

Nor was the detailed evidence as satisfactory as had been presented. Thus the fact that a crypt was only known about by the subject from York was not so surprising, since a very high proportion of churches of that date and geographical location would have had a crypt.

Such criticisms were made so strongly and effectively that the London *Sunday Mirror* serialization of the book published on the Bloxham tapes was stopped after the third of six installments. The presenter of a BBC program on the subject, Magnus Magnusson, retracted his earlier support for the cases on looking further into the situation.

The moral of all this is that the human brain is indeed a remarkable machine. Data can be stored in it for considerable periods, only to be produced in a much more coherent fashion at suitable times. This is an interpretation I will presently give of the cases of people who claim they are creating the music or paintings of long-dead great masters at the direct behest of the spirits of the latter. The creations are sometimes remarkably similar in style and could even deceive a supposed expert on the particular master's work. Conscious forgers can also do this, but this band of creative mediums claims to have had little talent or instruction in their métier before their spirit guide took command. Yet it may be that the sensitives might have been exposed to the works of a particular master at a tender age—though often this is almost impossible to discover. As we have already noted in Ryall's case, the human brain is a remarkable instrument of information storage. Cases of eidetic memory are known where a picture is stored in the brain without any conscious effort, and it can be returned to later and further details contained in it consciously noticed. Such memories, associated now with great compositions, might be overlaid by further ones as life continues. Thus it is not impossible that in later life the general style of a great creator could be emulated using these remotely glimpsed memories as a guide. Careful hypnosis might indicate this, though that need not be easy.

It is interesting to note that such a method has been used recently in various criminal investigations in America. For example, hypnosis was used in the case of the kidnapping of twenty-six schoolchildren by several men in Chowchilla, California. Ray, the driver of the bus from which they had been abducted, had seen the license plate of the van used in the kidnapping, but was completely unable to remember its number.

Hypnosis was then used on him. The police officer who was involved in the case remarked, "The important thing is that he wanted to be helpful. Hypnotic technique is most effective with either witnesses or victims who are eager to cooperate with the police. So with Ray, I got him to imagine he was watching television. I told him an instant replay of the abduction episode was about to run on the screen. It began, and when he described the point when the van was leaving with the kidnappers, I instructed him to zoom in on the license plate. His mind applied its own magnifying power to the plate and he was able to see once again the numbers on it. Then he read them off to me, digit by digit. Shortly thereafter, this information led to the arrest of three suspects, who will be tried for the crime." They were, indeed, the kidnappers.

There are other cases of a similar nature where hypnosis has allowed criminals to be found who would otherwise have got away. These reports show how remarkable a storage machine is the human mind, even though all that is stored is not necessarily conscious. Thus people like Edward Ryall are not in any way conscious of the deception they are practicing. Indeed, it has been perceptively put that "we all of us have a B-movie running through our heads."

There are many stories of haunted houses, some of which are very dramatic. However, accounts may have been added in their telling, so that reported cases should always be regarded with great caution.

There have been attempts to authenticate some of the hauntings that have been reported. Serious investigations of apparitions were made especially toward the end of the last century by the leading members of the British Society for Psychical Research (SPR). Many cases of hauntings were catalogued and witnesses to the more reliable ones interviewed further. A proportion of these cases involved apparitions of those simultaneously going through traumatic experiences, often that of death. Since telepathy could be invoked to explain such cases, we will not analyze them further here, except to say that the detailed evidence gained from them is not convincing. In many of the instances no evidence was available to prove that the percipient had remarked on the hallucination to someone else before hearing of the accident to the person whose apparition he had seen. Of the thirty-two cases reported by Edmund Gurney in *Phantasms of the Living,* published in 1886, only nine could produce any note made of the apparition at all, and none of those appeared satisfactory.

An analysis of "postmortem" ghosts—those of people who had been dead at least twelve hours before their ghost was seen by any living person—was made a little later by another SPR leader, Professor Sidgwick. She said that of the 370 cases in the SPR files at that time which "believers in ghosts would be apt to attribute to the agency of deceased human beings," the majority could be explained as illusions or hallucinations. There was, however, a residue which could not easily be explained in this way. The largest group of instances were those in which two or more people had seen, independently of each other and at different times, very similar apparitions. This included the traditional haunted house situation. These cases impressed the investigators of their time, but did not lead to strong support for the existence of spirits of the dead. For the ghosts were always reported as clothed. It is difficult to conceive of spirits in clothes.

Haunted houses are constantly being reported. One case, described to me by a friend, is typical. She arrived late one night at an old North Country hotel where she had booked a room. On entering, the room seemed hostile and she sensed that there was "something there," as she put it. She knew that she could not stay in the room another minute and went downstairs to get a drink and recover her equilibrium. She mentioned to the girl who served her that she would like to change her room. This was soon accomplished, two traveling salesmen who had imbibed quite freely being happy to make the switch. They were in no state to notice anything by the time they went to the room. The following morning my friend was served her breakfast by the same girl who had helped her the previous night. This girl explained that she had not wanted to say anything at that time but that she herself was terrified of that particular room and stayed in it for as short a time as possible whenever she had to clean it. Legend had it that a past owner of the house had killed his wife in the room about a hundred years ago.

One could explain this instance as caused by some unpleasant aspect of the room—its shape or decor, say—affecting those with more nervous sensibilities. Yet it is strange that both my friend and the girl described their very definite sensing of the presence of someone in the room with them.

Modern ideas of life after death have naturally attempted to keep pace with science. A recent believer in reincarnation writes, "The body of every individual has a physical and a supraphysical component; and when the energy exchange between these two components ceases to exist, the physical body dies. But the supraphysical body does not die. It cannot die: for the simple reason that it consists of an order of matter which is not subject to the process which we call 'death,' a process during which the physical particles integrated by an energy field have become inactive." This is an apparently authoritative statement backed up by the hard technology of modern physics—particles, fields and all. Yet such words are only as good as the evidence for them (as is also true in physics), so let us consider what form that evidence takes.

The various physical features accompanying ghosts—clothing and noises in particular—make it difficult to understand how the apparitions could be of spirits. Even if there were an immortal "supraphysical" body associated with our normal physical one, one would not expect it to be clothed nor to make the sounds associated with shoes on hard floors. The most natural explanation of such phenomena could be that of some "record" impressed on the surroundings by people involved in a traumatic experience such as murder. This was suggested by W. Myers in the late 1880s, though strongly attacked by another SPR member, Frank Podmore, who pointed out that few such cases have any link with some particular deceased person. Podmore also remarked that perceivers of such apparitions not infrequently had other hallucinations. Nor, he claimed, are similarities of visions of ghosts seen by different people as strong as all that; they rest rather on the absence of recorded differences.

New techniques to investigate hauntings have been developed which might prove of value in the future. Besides the original descriptive one of collecting all information possible from witnesses, on-the-spot investigations have been carried out with animals to observe their reactions to a supposedly haunted location. In one case a dog, a cat and a rattlesnake all reacted very strongly against an unoccupied chair in a haunted room in a Kentucky house, as reported by Graham Watkins. Another approach has been to take a consensus of "psychics" as to whether a house is haunted or not, and if so, whereabouts. However, nearly a century later there does not seem to be much that can be added to the problem of "postmortem apparitions." The evidence has not improved, nor has it modified the force of Podmore's remarks. It is not clear that there is anything truly paranormal happening when people see ghosts, above and beyond the experience of visual hallucinations at times of heightened sensibility. If there is anything more, then it need not be evidence for the existence of spirits. It could best be explained as caused by distortions left behind in the surroundings by people in stressful situations. These traces would excite visual and auditory hallucinations in the percipient by some type of "active radar" process. They might even cause physical effects of a poltergeist nature to occur, caused by the percipient.

Certain people appear to be able to act as intermediaries with disembodied spirits, either by voice (often changed to appear similar to that of the deceased) or by so-called "automatic writing." One of the most famous mediums of the last century was Mrs. Leonora Piper. Among the best examples of her ability was that which occurred with the visit of a man and wife whose little daughter Katherine (Katie) had died six weeks previously. Mrs. Piper spoke and acted in the way the little girl had done. She described her and gave her nickname, and told how she had died of a throat infection. She gave the names and nicknames of the little girl's brother and sister and of the little girl's favorite toys. As remarkable as the case is, it can still be explained naturally if Mrs. Piper had seen an obituary notice in a local paper. There might also have been clues given by the sitters—even a low murmur of the girl's nickname would have been enough. At the best it seems as if Mrs. Piper had shown telepathic powers. As Professor Sidgwick concluded of the spirit hypothesis in 1898, "He could not say more than that a prima facie case had been established for further investigation, keeping this hypothesis in view." At the worst, Mrs. Piper was particularly acute at picking up information from newspapers and the like and also noting nonverbal clues from the sitters.

One of the problems with such communications is the erroneous information sometimes forthcoming. I have myself been told quite incorrect facts about my grandparents by a medium. She then explained these "facts" away by saying they must be relevant to the other person in the room. When that also was found to be incorrect, the so-called information was ascribed by the medium to the presence of "evil spirits," who were warned off. On the other hand, information might have been given of which I had no knowledge but which could have been verified later. That was not the case here. However, it is not clear that I might not have such information somewhere in my brain but not available to conscious recall. Nonverbal cues might then have given it away to the sensitive eyes of the medium.

To establish the use of such cues it is essential that séances be tape- and video-recorded so that both sitter and medium can be observed with as much care as needed to ascertain the possible use of such cues.

Stories of the supernatural, especially about ghosts and like horrors, have large audiences who soak them up. The booksellers' shelves are full of books with covers and titles meant to curdle the blood of mere mortals. Some of these stories are supposedly based on true cases of haunted houses. But how real are the reports?

I would like to suggest that at least some of these are only slightly strange but are magnified out of all proportion in the mind of the believer. It is then very difficult to find out what actually did happen, since eyewitness accounts would have become too distorted.

I was involved recently in a case which could easily have been of this sort if it had not been nipped in the bud. Late one night I visited Borley church, close to Borley rectory, of which there had been many accounts of poltergeists and ghosts. With my colleague E. Balanovski, I listened inside the church with tape recorder and other equipment, all in silence and darkness. No strange effects occurred, though I heard a number of noises. However, all of these could be tracked down to natural causes. At one time the electricity meter in the church made some strange clicks; at another, bats in the church came close to my recorder and produced hisses, while a frightening whirring sound resulted from instrument malfunction. Previous investigators, some years before, had presented tape-recorder evidence for a ghostly visitation—footsteps, sighs and all! But these could all be explained quite naturally, at least to my satisfaction, from the sounds we had heard during our sojourn in the church.

The most startling occurrence of the evening took place when we were just packing up. It was the sudden appearance of the white-suited BBC director who had made the investigation possible. Having locked up the church, he had deposited the key at the vicar's house nearby and emerged suddenly out of the dark. It so happened that four young men were visiting the church at the same time on a "ghost-hunting spree," and they were petrified at the approaching white figure. Only my insistence that this was not a ghost—shouted at the fleeing backs of the young men— prevented a new addition being made to the stories about Borley church!

The phenomenon usually called dowsing (described briefly at the end of Chapter 6) is the practice of finding underground sources of water or other materials, such as oil or metallic objects, by using reactions of the human body to the presence of such material. Amplification of the dowsing response is often achieved by using a forked or Y-shaped twig held with one end in each hand and the forked end pointing forward. The dowser's hands are about body width apart, the forearms horizontal and palms upward. This means that the branches of the twig are bent outward. If the twig is in a horizontal plane it is easily liable to rotate so that it becomes vertical with the forked end pointing either up or down.

A dowsing search is performed by the dowser walking steadily across ground under which water or other objects are suspected of being buried. He commences with the twig held in the horizontal plane as described above. A dowsing reaction occurs at any point where the twig rotates in his hand, usually to point downward. The buried object is then claimed to be directly underneath. The size of the buried object, or its track if it is a stream, pipe or cable, can be ascertained by walking along various paths and plotting the points at which dowsing reactions occur. The depth of the object can be obtained by noting the strength of the dowsing reaction. Dowsing can also be done using two L-shaped rods held in cylinders, one in each hand, so that they can rotate freely. The long part of the "L" is held horizontal, the forearms again being held parallel and horizontally in front of the dowser's body. The dowsing reaction occurs when the rods turn in sharply to cross, or occasionally turn outward. It is also possible to dowse using a pendulum held in one hand, the dowsing reaction being the sudden change in direction of swing of the pendulum bob.

The use of dowsing goes back many centuries. The most ancient reference to it is a Chinese engraving from the first century A.D. showing the Emperor Yu holding a forked dowsing rod. Divining rods were used in the sixteenth century by miners searching for metal ores, though the practice was condemned by Martin Luther. In modern times dowsers have been employed by public authorities such as water, electricity and oil companies, and also by many private individuals in the search for water, oil and broken cables. Dowsers have claimed the ability to detect buried objects of archaeological value. Under the title of psychometry, dowsers have described the past history of particular objects given to them. By means of dowsing it has even been claimed that medical diagnosis is possible.

One example of the diagnostic power of a dowser, and clearly related to the supposed power of finding buried metal, occurred several years ago.

On that day the sun shone thinly on a Welsh valley, with the river rushing by a hundred yards away from a standing stone. A line of sticks had been put in the ground leading away from the stone, being the path of a deep underground stream according to the Welsh dowser Bill Lewis. He also claimed that another stream ran beneath the first to cross it exactly where the standing stone had been sited. Bill could even give values for the depths of the streams as well as their flows. All these claims were very hard indeed to check, yet various people had sunk wells successfully at places chosen by Bill in their search for water. Bill Lewis used a pendulum to indicate where the water was situated.

A television interviewer visited Bill Lewis as part of a TV program on the paranormal. Mr. Lewis gave a diagnosis of the interviewer's state of health which included the puzzling claim that there was a piece of metal in the interviewer's thigh. Such a diagnosis was not very likely to be true. Yet indeed it was so, the interviewer having had an operation requiring the reinforcement of that particular bone by a piece of metal. It would have been difficult (though not completely impossible) for Mr. Lewis to have discovered the fact about the embedded metal beforehand, without having a very efficient spy network. Nor was there any defect in the interviewer's limb movements caused by this metal insertion.

However, I now regard this claim by Bill Lewis as a coincidence.

There is data on success of dowsed versus nondowsed wells from two sources. One is from a community called Fence Lake in New Mexico. Out of a total of sixty-two wells drilled, there were twenty-four successful wells which were dowsed as against twenty-five that were not dowsed, while five dowsed wells were dry as compared to seven dry nondowsed wells. A much larger set of cases has been collected from central New South Wales in Australia over the period 1918 to 1943. Out of a total of 3,600 wells drilled for water during that period, 70 percent of the dowsed wells produced over a hundred gallons an hour as compared to 84 percent of the nondowsed wells. There were twice as many complete failures in dowsed wells (14.7 percent) versus the nondowsed wells (7.4 percent), although this difference may have been due to water diviners being called in to areas with less underground water than in the nondowsing cases.

These statistics are strong evidence that the dowsing reaction is not correlated with the presence of underground water. There is no similar evidence in regard to dowsing for oil or minerals, so we cannot make any rigorous conclusions about that. We need to turn to laboratory tests of dowsing for such materials and others to see if there is any possibility of a paranormal phenomenon being involved.

A very careful study of the dowsing faculty was carried out during the first week of August 1949 by the American Society of Dowsers. A field was carefully chosen so as to have as few visual clues for the presence of water as possible. Twenty-seven dowsers each selected the best spot to sink a well, and estimated the depth of the well and the amount of water it should contain. A geologist and a water engineer also made estimates of the depth and rate of flow of the underground water at sixteen different places which had been previously chosen. Wells were then sunk at the various points advised by the dowsers. The two experts proved quite successful at estimating the depth of underground water. The dowsers were a complete failure. As the authors wrote: "Not one of our dowsers could for a moment be mistaken for an 'expert.' . . . We saw nothing to challenge the prevailing view that we are dealing with unconscious muscular activity, or what Frederick Myers called 'motor automatism.' "

A careful series of tests were performed at the Military Engineering Experimental Establishment of the British Ministry of Defence. These were to determine how successful were map and on-site dowsing for buried mines (where in map dowsing the position of the desired object is marked on a map of the location by the dowser even though he may be distant from the site). Positive results would naturally have proved of value to the particular establishment, but dependence on nonexistent powers to detect buried mines could clearly have a fatal outcome.

Map dowsing was tested by burying twenty inert mines at points on the 6.7 miles of roads and tracks at one of the establishment outstations, a map at a scale of 1:2,500 being given to the dowsers. Seven dowsers made various attempts, but none was better than chance at any time, even when one of the dowsers had practiced for a period of time with a sample mine. Members of the establishment also attempted to guess where the mines were, but again the results were no better than expected by chance. The experimenter concluded: "To sum up on map dowsing, the results from all the trials are really failures and there is no evidence that this is a practical method for locating mines."

On-site dowsing tests were then carried out, using a four-hundred-foot square area of heath and heather divided into two hundred twenty-foot squares. The dowsers' ability was tested to see if they could distinguish between buried metallic or plastic mines, concrete or wooden blocks or nothing buried at all; the concrete and wooden blocks were the same shape and size as the mines. Nineteen dowsers covered the full course, but only one dowser gave results of reasonable significance, and only then with a high effort in digging many holes not containing mines. The conclusion, again, was that there was no real faculty possessed by certain people allowing them to detect buried objects of various sorts.

Finally, water divining was tested, using buried pipes which might or might not be carrying water. One experienced dowser was asked to say whether or not water was flowing in a single pipe, which had water flowing in it or not according to a random sequence unknown to the dowser. He was right exactly 50 percent of the time in fifty attempts, exactly according to chance. There were also tests on four junior officers who had supposedly been trained by another experienced dowser, again giving no results beyond the chance level.

We have thus reached the conclusion that there is no strong evidence whatsoever to support the existence of a dowsing faculty of any sort. All carefully controlled tests of people who have claimed to possess dowsing powers have shown such faculties to be conspicuously absent at the time of the test. The dowsing reaction itself, in which the twig suddenly twists in the dowser's hands or the pendulum changes its direction of oscillation, can best be explained as an unconscious muscular reaction to some sensory clue. The usual claim of the dowser that he has no control over the twig is true if he means conscious control, but need not be so if unconscious mechanisms are included.

But in any event it is also necessary to take account of any possible "clues." When they also are taken into consideration, even the anecdotal case of dowsing need not appear paranormal at all.

The apparent ability to observe phenomena at a distance, used by a number of people to discover the whereabouts of missing persons, may be explicable in the same way. One of the best known of such experts, Gerard Croiset of Utrecht, discovered the body of a child, Lesley Ann Downey, missing since the previous Christmas in the famous "Moors murder case." Her mother had written to a London magazine seeking Croiset's address. Instead of just giving it, they decided to sponsor his investigation. Some of her toys were flown to Holland and a week later Croiset came to Manchester, having been unable to trace the child from such a distance. Police and detectives involved in the case gave him every facility and within an hour of his arrival he announced that the child was dead and her body buried in a shallow grave near a large expanse of water. He directed the car in which he was a passenger to a number of places where, he said, the child had been driven by her two abductors. He added that she had entered a large building from which she did not emerge alive. The following December Ian Brady and Myra Hindley were charged with the murder of two boys and Lesley Ann Downey, whom Croiset had tried to find. During the trial at Chester in 1966 evidence showed that she had been driven to the places indicated by Croiset, murdered in a block of flats and subsequently buried in a shallow grave on Saddleworth Moor near a large reservoir.

This is only one of a number of cases in which Croiset has apparently been successful. One case is even more remarkable, in that Croiset indicated where the body of a little Japanese girl, who had been missing from her suburban Tokyo home, would be discovered. The following morning a television crew went to the spot, following Croiset's directions, and found the girl's body exactly as Croiset had described. The police had been searching in a completely different area for the preceding weeks.

It is very unlikely that Croiset could have used a spy network to achieve his results, especially in the Moors murder case or that of the little Japanese girl who had disappeared. But we have to assess the probability of Croiset making a guess as to where the girls were. Thus, in the Japanese girl's case it would seem reasonable to assume that the little girl could have strayed and fallen into a nearby lake. Croiset himself is concerned with death by water since he himself nearly drowned as a young man. Yet there are features which make some of the cases exceptional. Descriptions of scenes are given which are almost impossible to conceive as having been obtained by chance; this occurred in both the Moors murder and the Tokyo case.

Another example of this is the case of Pat McAdam, a girl who disappeared from her home in Scotland in 1969. Croiset was asked to help in her discovery after newspaper appeals and police searches had failed to turn up any clues as to her whereabouts, either alive or dead. Croiset drew various pictures of sites where she had been just before her supposed death by violence and the subsequent disappearance of her body along a river and out to sea. He gave remarkably accurate descriptions of certain places where she could well have been taken by an abductor.

One place was supposed to be a bridge with iron railings across a stream flowing through a narrow valley. On either side of the stream, tree roots protruded; on one side of the stream was a newly painted house with a sign which indicated that it had been involved in some business activity but was no longer so. There is indeed such a spot on a road which the girl might have crossed before her death. The stream is bridged in exactly the way described by Croiset, and had a house next to it which had a sign indicating that it had been used as a real estate office.

The girl's body was not found, though Croiset claimed that some of her belongings would be discovered at a bend in the river where her body had been entangled in bushes. Various articles belonging to a woman were indeed discovered which corresponded to items she might have bought during a shopping expedition just before her death. Definite identification of these objects proved impossible due to lack of local police cooperation, however. In spite of that, the McAdam case must be regarded very highly on the grounds of the near impossibility of Croiset having been able to have guessed the various features. Using the pictures he drew, I was able to estimate very roughly the likelihood of his descriptions being right purely by chance; the result was much less than one in a million. Only through the use of an accomplice could he have been able to obtain such accurate information about distant scenes. Having visited Croiset's house and seen his mode of operation, it appears unlikely that he has a network of accomplices throughout the world. Since only one accomplice would be needed for the McAdam case or any other separate case, then each case alone cannot be used as validation. Only if all Croiset's cases are compounded is there less likelihood of a network of accomplices.

Yet not all of Croiset's attempts are successful. Even those which appear to be so may still be coincidental. Therefore, even though the chance of his guessing various features of the Mc-Adam case was less than a million to one, such odds still allow such unlikely things to happen, as in the case of a newspaper vendor, Mr. Stanbridge, to be described shortly.

We can only conclude that results like Croiset's may merely arise as fantasies in the mind of the medium, the coincidentally correct features then being seized upon and taken as evidence for the correctness of the whole. The same lack of feasibility of the phenomena, as in that of clairvoyance, leads me to suspect that such is indeed the correct explanation.

To summarize, we have considered in this chapter five types of paranormal phenomena: reincarnation, hauntings, messages from the dead, dowsing and distant viewing. We have seen that none of the evidence for them is strong enough to say that they are truly paranormal. According to the title of this chapter, they have all been classified as arising from cues, fantasy and memory. To see that this is so let us discuss the evidence presented.

"Reincarnation" was observed by regressing a subject under hypnosis. It involved memories of appropriate events used in building up the fantasy "earlier life" and released under the powerful influence of the hypnotist. The Edward Ryall/John Fletcher case is clearly similar, though without the need for hypnosis to uncover earlier memories (though some sort of "self-hypnosis" may have been involved).

The analysis of hauntings by the Society for Psychical Research showed they mainly involved hallucinations, possibly brought about by normal effects but when the subject was in a suggestible state; a clear case of this was almost caused by the ESP investigations of a colleague of the author! Nor did there seem to be more reality in messages from the dead than that provided by (possibly nonverbal) cues given to a "sensitive" by the subject receiving the message.

Dowsing was shown to have no solid evidence at all for its existence. Any of the various isolated successes could well be accounted for by perception of geological clues as to the whereabouts of water by the heightened sensitivity of the dowser; this explains the lack of success of dowsers when all possible clues are removed in careful tests.

Finally, the surprising successes of "distant viewers" like Croiset have to be set against their failures, as well as the suggestibility of some of those involved in such cases. The descriptions of the distant scene appear, on retrospect, to be more the fantasies of an artist (albeit expert at where, for example, peoples' bodies might be found after an accident) than the result of real paranormal powers.

11

The Framework of the Paranormal: Coincidence, Credulity and the Fear of Death

FROM 1926 TO 1937 GEORGE HENRY STANBRIDGE HAD RUN A newsstand at Archway Tube Station in North London. In 1958 he emigrated to Australia but returned to London with his wife in 1977 for a holiday. He happened to go to a cinema where a film called *The Hand of Fate* was playing. He was astounded to see on the screen a news vendor called George Henry Stanbridge, looking very much like himself and standing on the very same spot he himself had occupied for years. He was in for an even greater shock when he discovered that his screen "double" was supposed to have brutally murdered his wife during the Second World War. Mr. Stanbridge sued the distributors of the film and they agreed to withdraw it. They had had no knowledge of the real Mr. Stanbridge or that he had run the newsstand at Archway.

An amazing coincidence indeed; as Mr. Stanbridge said, "It was like taking a jump and landing up on the moon—a three million to one coincidence." The unlikeliness of such a coincidence actually happening makes other explanations more attractive—in particular that of the existence of some sort of direct (though unconscious) communication between George Henry Stanbridge and the scriptwriter of the film. There are many more direct cases reported of apparent thought transference which would seem to support this latter alternative in preference to a highly improbable coincidence.

There is the one reported to me recently of a middle-aged woman who told her husband one afternoon that they must immediately go to see her mother; the woman was sure that something was wrong with her aged parent, who lived alone. Since the mother did not have a telephone, a fifty-mile drive was therefore necessary. The husband agreed to make the journey because his wife had never acted in this way before and he felt that there might indeed be something amiss with his mother-in-law. On arriving at the flat where the latter lived, there was no answer to their insistent knocking. After much effort they managed to persuade the caretaker of the flats to use his skeleton key. On opening the door of the flat the woman found her mother on the floor in a coma. Fortunately she revived after being taken to hospital.

There are many, many events of this sort which are difficult to explain merely as coincidence. Nor is it easy to use the same grounds for dismissing premonitions of the future. A famous example of this sort (in fact one of several) occurred to Sir Winston Churchill in the Second World War, and was recorded by his wife. It was connected with his habit of going out in a car at night during air raids in order to boost the morale of London's civil defense forces. Sir Winston usually sat on the driver's side of the car. On one particular night he refused to enter the car through the door held open by the driver on his usual side, but went around and sat down in the other side of the car. As they were driving along the Kingston bypass at 60 m.p.h., a bomb fell near the passenger's side of the car, lifting it up onto the two opposite side wheels. Miraculously the car righted itself and they continued; if it had overturned undoubtedly Churchill and his driver would have been seriously injured, if not killed. As his wife reports, "He did not tell his wife so as not to scare her, but she heard about it from the driver, and decided to challenge him about the incident: 'Winston, why did you get in on the other side of the car?' 'I don't know, I don't know,' Winston answered at first, but his wife pierced him with her gaze and he realized he could not get away with that answer, so he said, 'Yes I do know. When I got to the door held open for me, something in me said "Stop, go around to the other side and get in there," and that is what I did.'"

A more precise form of premonition, though with a longer interval between the premonition and its actuality, is reported by Mrs. Louisa Rhine (J. B. Rhine's wife), who investigated many similar cases. A mother had a waking "picture" of her elder son, Herbert, dead in the bathtub. It haunted her so strongly that she made a special point of listening to check that nothing was wrong; she did not tell Herbert but only her younger son. Herbert then left home, but when he came back for a holiday she remembered the premonition. One evening she heard him whistling and singing in the bath. She was dressed to go out but could not leave. Then she heard the water running but now Herbert was silent. The mother finally opened the door to discover him lying exactly as she had pictured him in her premonition two years earlier. Her son had been overcome by fumes from the gas heater. She opened the door and the windows and called a doctor who revived Herbert. Without her watchfulness her son would undoubtedly have died.

Such a premonition is not unusual, according to a recent poll of American university students at a college in Georgia. An astonishing 65 percent of the 433 students questioned said they had had dreams that had come true; 23 percent even reported that it had happened more than once. Some of the dreams that came true included a bank robbery and family deaths, and one student even dreamed that George Wallace would be shot the night before it happened. Most of the dreams dealt with family members or close friends; the events in a dream often occurred within a week of the dream. About a tenth that came true involved death, accidents or sudden illness.

In a previous chapter I mentioned the reason for my frequent "premonition" that I was about to receive a telephone call. I cannot really consider myself psychic because I also am telephoned sometimes by someone I am just about to ring myself. I do not know the odds on this occurring by chance, but there are many situations in which the odds are higher than one would naively expect. For example, the odds that at least two people out of twenty-three have the same month and day for their birthdays is a little greater than one in two.

What of dream premonitions, such as the one described concerning the Moorgate tube disaster? First we will do a simple sum. It is known that every person has about seven dreams per night, and it is reasonable to suppose that one of these is very likely a nightmare of some sort. In other words there are expected to be about fifty million nightmares per night in England, two hundred million in America. Even if this estimate is too high by a factor of ten, there should still be hundreds of thousands of nightmares per night in most countries of the size of England. A proportion of these will be remembered on waking. If one then considers the variety of accidents that could have arisen in these dreams, it seems quite probable that some of the nightmares would actually come true over the succeeding days or weeks.

The Moorgate tube disaster premonition could indeed be of that sort, the nightmare possibly being sparked off by remembrance of the trauma of birth. This may also explain the Aberfan premonitions.

It is necessary to consider different explanations for the successes of prophets like Jeane Dixon. The famous assassination case is not so difficult to assess, there being a nontrivial probability of assassination of an American president. Indeed, all such cases of prophecy have to be assessed on how likely they are to be correct by chance.

This is comparatively easy in situations where the prophecy is of a rather general form, or it is not specified exactly when the event will occur. One particular prophet who claimed powers over the weather offered to bring rain to England during the great drought in the summer of 1976. His offer was made at the end of May, but no rain occurred until the end of August—three months later. The prophet claimed that this delay was caused by lack of authorization from the English government for the project of rainmaking. But that never came, even by the end of August, although the rain certainly did!

W. E. Cox, one of J. B. Rhine's colleagues, ascertained that there were significantly fewer passengers aboard twenty-eight railroad trains which were involved in accidents than on trains running at similar times a day or week before or after. The results were claimed by Cox only to have been explicable by chance once in a hundred times. Cox suggested that potential train passengers were aware of the oncoming tragedies but not at a fully conscious level.

Let us consider the accident-avoidance displayed by passengers on trains. A possible nonparanormal explanation of the facts reported by Cox is that train accidents tend to happen on days of bad weather. People will also tend to travel less on such days, because it is unpleasant, and even physically difficult if there is snow around. This simple explanation apparently has not been checked nor even mentioned by Cox in his reports on the subject. There may be other explanations of a natural form to consider, such as the possibility of a higher incidence of flu or colds on the days in question. It would seem necessary to look at detailed reports on the accidents before Cox's explanation in terms of precognition can be accepted.

This is also true of the spontaneous case experienced by Cox himself, and described in Chapter 2. He admitted, in a letter written to me, that "individual cases per se are not sufficient as proof." He also added in his letter that, in the case he quoted, "I told to none but my wife for two years." Two years is a considerable length of time over which to recall detailed events with great accuracy. This case should be treated like any others of a similar sort: unless there are reliable witnesses of the precognition at the time it occurred, then the evidence is not really acceptable. Until such is presented, the instance could only be interpreted as one of poor memory or of a misheard newscast.

Coincidences do occur at a rate expected by chance, even the very unlikely and unhappy one experienced by George Henry Stanbridge mentioned at the beginning of this chapter.

Life expectancy has about doubled over the past millennia; some people even live for over a hundred years. But death expectancy has remained unchanged. The process of aging has not yet been understood, and even when it is there is no assurance that it will dramatically increase life expectancy, let alone grant immortality. Naturally enough, man has hoped that this life is not all there is to existence. He may fantasize that he has many other lives both in the past and to come. Alternatively he may wish that he will be transformed at death so as to persist eternally in some new state. Such aspirations have been enshrined in religious beliefs of many kinds, some of which still hold strong attraction for men and women today. Thus Christianity includes as one of its tenets a belief in the personal immortality of human souls. Personal immortality is also one of the tenets of Christianity's sister religion, Islam.

However, the spread of disbelief, particularly in the West since the seventeenth century, has led to a concomitant increase in the fear of death. Indeed, the very word itself has become almost taboo, particularly in the United States. For example, there is a reluctance to tell a person that he or she is dying. An increasingly materialistic attitude has also developed among Christian theologians, with less discussion of immortality. These have all combined to leave modern Western man to fend for himself when he faces his own death. No wonder that there is an avid reading public for writers who claim that they have proof of survival after death, or that there is great interest in communications from the dead.

A number of people who have been on the point of death but have survived do give interesting accounts of their experiences. Many suffer out-of-the-body experiences. Thus a woman with a heart attack seemed to drift up from her body until she floated just below the ceiling. She reported, "I saw them below beating on my chest and rubbing my arms and legs. I thought 'Why are they going to so much trouble? I'm just fine now.' " There are those who may encounter intelligences. One man said, "I knew I was dying and there was nothing I could do about it. I was out of my body, because I could see my own body there on the operating table. All this made me feel very bad at first, but then this really bright light came. It was just too much light. And it gave off heat to me; I felt a warm sensation. Then it kind of asked me if I was ready to die. The light was talking to me, but in a voice." In the near-death experience there may also be visions of loved ones who have already died.

An American doctor, R. A. Moody, has talked to dozens of people who have nearly died. A composite of all the stories Dr. Moody heard would apparently go something like this: "A man is dying and as he reaches the point of greatest physical distress he hears himself pronounced dead by his doctor. He begins to hear a loud ringing or buzzing and at the same time feels himself moving very rapidly through a long dark tunnel. After this he suddenly finds himself outside of his own physical body and he sees his own body from a distance, as if he is a spectator." It is then that the person may begin to glimpse spirits of relatives and friends who have already died, and "a loving, warm spirit of a kind he has never encountered before—a being of light—appears before him. . . ."

These accounts have been confirmed by Dr. Kübler-Ross from her own work with terminally ill patients. She writes, "It is research such as Dr. Moody presents . . . that will enlighten many and will confirm what we have been taught for two thousand years—that there is life after death." She also says about her own work, "From my interviews with the dying and with mediums, I would describe the other world as looking similar to ours except for the colors, which are very vibrant. We will be met by the ones we loved the most in this life. I have hundreds of times seen dying patients speak to people who died earlier. Dying people all say they are met by deceased loved ones. If they have been mutilated in life they feel whole again. On the other side you see whatever you have loved on this side."

The experiences reported above are not used by Dr. Moody himself to prove that there is life after death, yet they are regarded by Dr. Moody as strongly in support of it. There is no question, at least to Dr. Moody, that the accounts are true. "The notion that these accounts might be fabricated is utterly untenable," he claims.

Such experiences are similar in content to those involved in LSD trips or in sensory-deprivation studies. They may also occur in other out-of-the-body experiences. The details of the experiences are interesting in themselves. However, they give no evidence for the independent existence of the intelligences encountered by those experiencing the altered state of consciousness. They cannot therefore be used in any argument for survival after death in any form.

Indeed, there are various explanations of these experiences according to orthodox science. A consultant anesthetist in an English hospital who has been involved with many organ trans-

plants and had a particular interest in the circumstances sur-
rounding death, has described the orthodox medical explanations of such experiences. The possible explanations for near-
death visions could be grouped into four main classes. The first
is pharmacological, the effects arising from drugs often adminis-
tered to people on the verge of death. There are certain drugs,
especially kitmomine, which produce out-of-the-body sensations.
The second is neurological, malfunction of the brain near death
causing hallucinations. The third is physiological, the visions
arising from lack of oxygen or reduction of blood sugar. There
is finally the psychological explanation, due to possible sensory
deprivation as death approaches.

Put together with the analysis in the previous chapter, the
evidence for life after death does not hold up. At best it is suggestive; at worst downright misleading. All the cases I have investigated have turned out to prove just the opposite to what they
were supposed to. There is therefore no real need to go through
such a discussion as in the previous chapter. Yet we should not
neglect analyzing the possibility of life after death. It is only right
that modern scientific results should be brought to bear on this
question which has perplexed and worried so many. We shall
have to face death sooner or later, so that its relevance to our
human condition will never cease.

There are two possible "mechanisms" for a putative life after
death. One involves the continuation of physical energies which
have some relation to the living human from which they arose.
The other assumes the existence of a nonphysical "soul" or
"etheric body" which carries the imprints of its earlier bodily
form. These are not necessarily mutually incompatible, since
there may be continuation of a human being in a partly physical
and partly nonphysical form.

It is also necessary to consider the other possibility of avoid-
ance of the complete destruction of a human personality at death
—and that is reincarnation. However, with such an idea, the same
problem has to be faced as for life after death. Does the residual
personality at death persist in a physical state or in a nonphysical
one until it is associated with another human body? It is possible
to propose that the nature of such existence need not be a diffi-

culty if there is an instantaneous transfer of the personality residue from a person at death to one just being born.

This suggestion has several serious defects. The first is that simultaneity is not well defined in modern physics. What is simultaneous to one observer is not so to another. Nor is it clear precisely when this personality residue should leave a dying person. What of someone who is rescued from death after being in a coma for a period of time? Have they hung on to their immortal residue after all, even though at times they were thought to be dead? And when is it the right time for this residue to enter its next body? At conception? A month after conception? At birth? We conclude that since simultaneity is not well defined and because of the other reasons given, reincarnation has to be considered as involving two possibilities for the existence of the immortal residue after its body's death—either a physical or a nonphysical one.

The second of these two possibilities takes us outside the materialistic framework which was strongly argued for in Chapter 3. We showed there that a nonphysical approach, not being quantifiable, was therefore untestable. This latter feature was so even for the holder of ideas developed in a nonmaterialistic framework. It is therefore impossible for such ideas to be discernible from nonsense.

This conclusion will undoubtedly incense all those who believe in "spirits" or "etheric bodies" or the like. I can only ask them to be more precise about the meaning of these terms. When I ask for clarification of this sort, I am usually told man cannot understand such questions. If so, then how can he even know that he himself is talking sense when he uses the "concepts"? Thus, the Christian theologian, Dr. Leslie Weatherhead, states, "A spirit doesn't occupy space and an etheric body doesn't." Yet he had stated earlier, "If you had instruments fast enough and sensitive enough you could recover that light (man's consciousness). It's still vibrating through the universe after the candle itself is burnt out." But then the soul or spirit must have size, indeed that of the whole universe itself. Dr. Weatherhead wishes to have it both ways—the soul has size, but yet it does not have size.

I have said enough against the idea of a nonphysical immortal soul to enable me to reject it as impossible to use in any rational and scientific way. It will always appeal to some, but then belief is itself irrational. Let us turn, with hope of more clarification, to the problem of a physical soul. Can there persist some sort of energy which would accommodate enough of a person's personality after the destruction of their body to allow some form of immortality to be achieved?

There have been recent developments in physics, both of a theoretical and experimental nature which, at least on face value, give the hope that this might be possible. They go under the name of "solitons," which are persistent packets of energy that can float around and even pass through each other without much harm. The first published account of such a possibility appeared in 1844 and was given by Scott-Russell: "I was observing the motion of a boat which was rapidly drawn along a narrow channel by a pair of horses, when the boat suddenly stopped—not so the mass of water in the channel which it had put in motion; it accumulated round the prow of the vessel in a state of violent agitation, then suddenly leaving it behind, rolled forward with great velocity, assuming the form of a large solitary elevation, a rounded, smooth and well-defined heap of water, which continued to course along the channel apparently without change of form or diminution of speed. I followed it on horseback, and overtook it still rolling on at a rate of some eight or nine miles an hour, preserving its original figure some thirty feet long and a foot to a foot and a half in height. Its height gradually diminished, and after a chase of one or two miles I lost it in the windings of the channel. Such, in the month of August 1834, was my first chance interview with that singular and beautiful phenomenon."

Such solitary waves, or solitons, have been observed in other situations besides water waves, such as in ion-acoustic plasma waves and in nonlinear crystal experiments. It has even been suggested that the red spot of Jupiter is a soliton of the planetary atmosphere.

We therefore consider the possibility that the immortal soul is a soliton of the energy fields of the body. Our discussion in Chapter 3 on the fields relevant to human activity indicate that the only possible field is that of electromagnetism. We should also note that only a soliton-type of model of the soul is possible; persistence of the packet of energy and information is absolutely essential for true immortality.

We have already mentioned a case in which solitons occur in the electromagnetic situation, namely for ion-acoustic plasma waves. But the plasma, or gas of charged particles, is a very different physical system from that which occurs in the human body. There have been claims from Russian parapsychologists of the existence of a "bioplasma" energy associated with the human body. However, bioplasma has only been a word attached to phenomena such as the "Kirlian effect," which our investigations have shown are of little relevance to the human condition.

There does exist in nature what is very likely an electromagnetic soliton of great interest to us called ball lightning. As to be expected, it appears as a brilliant ball of light, a foot or so across, usually during lightning storms. Reports indicate that it can persist for minutes. Yet it can be produced apparently only by strong electric fields, usually in thunderstorms. Fields of the order of millions of volts per meter arise in such situations, so that similar field strengths would seem to be needed to produce them artificially.

The only evidence we have from nature of electromagnetic solitons are those created in fields and charge densities at least thousands of times greater than that conceivable for human beings unaided by special machines. In any case the information content of such objects would appear to be very low indeed. The possibility of them containing enough complexity to describe anywhere near the personality of a living human being is very remote. They are essentially balls of fire, with a very simple structure. In spite of its deceptively smooth structure, "like two fistfuls of porridge," the brain is extremely complex due to its vast number of interconnecting neurons, each of which is also a highly complicated machine.

Let us look in a little more detail at the possibility of the existence of a single soliton of matter and radio-wave energy. Unless it involved shortwave energy and compressed matter it would rapidly disperse, like a cloud. We have indicated that there is no evidence for such shortwave energy emission in humans other than the natural "black-body" radiation. Such low levels of radiation as have been observed would in no way hold the matter together; it would again disperse. We can only say that there is no chance of the soul being a matter radio-wave soliton.

So we come to the realization that there can be no immortality in a physical world. In a nonphysical universe, anything is possible, but since we cannot be sure that anything makes sense, we cannot consider it usefully. According to modern science there is only death after death.

There is a high level of gullibility where seemingly inexplicable events are concerned. Even apparently sober people such as policemen and airline pilots appear to become very naive after having witnessed something in the heavens which they cannot readily explain.

To take an example, a police sergeant in Kent witnessed what was an "unidentified flying object" flying overhead at four hundred feet. I questioned him as to how he knew it was four hundred feet. "Because that's the usual height planes fly over here," he replied, with absolutely no further reason. How on earth he could expect his UFO to fly over at the same height as a plane I still do not understand. This height of four hundred feet was crucial to his estimates of the size and speed of the UFO itself, and consequently of its abnormality.

Another case involved the captain of a plane who, together with his crew, saw a strange light high in the sky just before dusk on one flight; they noticed radar reflections from the same direction on their return flight several hours later. Some months later I questioned the captain's claim that the light was from an "internal source" and could not have been a noctilucent cloud acting as a mirror to the sun as it went down. His response was strangely violent, as it was to my suggestion that the radar signals might have been "mock" or false echoes. I soon discovered that he had firmly believed in UFOs since the experience nine months previously. Yet at no time during the intervening period had he investigated the quite extensive literature available on the nature of noctilucent clouds or radar echoes. I am afraid that here again the evidence for UFOs is so slight that no serious claim can be made for them. Intelligent extraterrestrial life may well exist, but there is no real evidence that it has ever visited our earth.

Even more trivial is the Bermuda triangle phenomenon. I quote with approbation the English journalist Brian Inglis's recent comments on the phenomenon: Being allergic to such mysteries I did not read *The Bermuda Triangle* when it came out a couple of years ago. But I read some of the criticisms and I have been interested to see how Charles Berlitz handles them in its successor *Without a Trace*. The author of a book which sells over five million copies, after all, has something to answer for if the critics have been able to make damaging allegations about the way in which he handled his material. As in the foreword to the new book he asks rhetorically, "What is the purpose of another book about the Bermuda triangle?" it is only reasonable to assume that one of the purposes will be to show that the critics were mistaken. But this is not Berlitz's way. He adds fresh evidence; but he says blandly that "the aim of the present book is neither to refute, inform or educate the critics." And this relieves me of a task which I had thought might be imposed; to read *The Bermuda Triangle,* as well as its successor, in order to judge dispassionately between the critics and Berlitz. It may well be that there is more to the triangle than they conceded, but if Berlitz does not care to answer them, there is little point in proceeding any further.

Hear, hear!

There are many anecdotal reports in the business community that executives do not think logically when they come to a decision. Alfred P. Sloan, once president of the American firm General Motors, summed up the attitude of thousands of executives by saying, "The final act of business judgment is intuitive." Sloan was himself characterized by W. C. Durant, the founder of General Motors, as "a man who could proceed on a course of action guided solely, as far as I could tell, by some intuitive flash of brilliance. He never felt obliged to make an engineering hunt for the facts." A former chairman of the board of U.S. Steel said about decision making: "You don't know how you do it; you just do it." An executive vice-president of Bethlehem Steel described this in the same manner: "Intuition is nine-tenths of my job. In my company we have over a hundred specialists. If I need any information I go to the specialist and get it. But I mustn't spend too much time with him, or else I'd become a specialist too. And that's the thing I must guard against in every way I know how. If I can remain the generalist, intuition will keep on flowing through me and I'll make the right decisions for my company— even though I don't know how I do it."

To test if business executives were using a precognitive power, American researchers asked each executive to guess a chance sequence of one hundred 0's or 1's before it was produced by a computer, one for each particular executive. A score of more than fifty correct 0's or 1's indicated above-chance success. In this way it was possible to assess if a person could predict the future. At the same time each subject was asked to pick a metaphor to be used to classify their "dynamism." The metaphors involved a description of the subjects' attitude to time at five different levels, ranging from the most passive "oceanic" one to the most active "dynamic" one.

The tests were claimed to be successful. In seventeen out of the twenty-two groups of business executives a positive relationship of dynamic to high-average score or oceanic to low-average score was found. Only eleven out of these twenty-two groups would have been expected to give such a result; the odds of this occurring by chance were found to be one in a hundred.

I also tested for precognitive ability among two groups of business executives, about thirty in each group, along the same lines. After a talk on various aspects of intuition and its possible paranormal basis, I asked each member of the group to write down five lines, each with ten 0's or 1's in them. I asked them also to note down their attitude to time (either "dynamic" or "oceanic," only the two extremes of the American test) and whether or not they believed in ESP.

About 80 percent believed in the existence of ESP and it was clear from later discussions that such belief was very firm indeed in some of the executives. But the real result of the test came when I proceeded to generate for each of their fifty numbers a similar number purely by chance (flipping a coin and assigning 0 to heads, 1 to tails). Each number so generated was compared with the number written down earlier by the executive, and scored according to the agreement between the numbers. A high score (much above twenty-five) for any person could be interpreted on the grounds that they had precognized the number I had randomly generated.

The result of my analysis was that there was no significant difference between the scores actually achieved and chance level. However hard I "played around" with the data, I could not come up with any significant feature beyond chance. None of the executives seemed to have "paranormal" powers. I could have altered the data a little—and then obtained slightly suggestive results. But in any event statistical analysis has to be treated with a great deal of caution.

One might well ask how a band of parapsychologists, some of them with Ph.D.'s in physics, chemistry or mathematics, could persuade themselves that the absolute contrary to their science is actually occurring? I almost did so myself! I think the answer has already been given by the archsceptic, James Randi, to his own question as to how Puthoff and Targ could have allowed such loose conditions in earlier ESP tests at SRI with Uri Geller, such as submitting to Geller's demand to be let out of the sealed room? The answer was: "Because I have seen what grown men will do to satisfy a deep need to believe." I agree wholeheartedly with that remark.

In summary, we have shown in this chapter that premonitions of the future, survival after death, UFOs and the Bermuda triangle are all phenomena happening by sheer chance (in the case of premonitions of the future) or due to natural causes (in out-of-the-body experiences related to survival after death, UFOs and the Bermuda triangle disappearances), playing on man's credulity and his fear of death. Exaggerated or distorted evidence, and inability to appreciate the role chance or nature can play, have led many to give great importance to these events. But the vast wealth of evidence for the various phenomena collapses to nothing when looked at with more care. We learn again the lesson gained earlier from more careful analysis of other paranormal phenomena: on more careful investigation the supernatural disappears; what is left is wholly natural.

12

Unresolved Problems

OUR INVESTIGATIONS ARE FINALLY OVER. WE HAVE SEARCHED for the supernatural and not found it. In the main, only poor experimentation, shoddy theory and human gullibility have been encountered. There is also the realization that nearly all of the claimed paranormal phenomena are in complete contradiction to established science. We have constantly tried to take the line that the phenomena should not therefore be ignored, that such a line could only stem from an attitude of utter closed-mindedness, which augurs badly for any attempt to understand the world.

Having admitted that, we must honestly declare that there are those psychic phenomena which are still difficult to explain completely and finally. These are of two sorts, those depending on the reports of others and those which I have witnessed with my own eyes.

Of the first, we can classify spontaneous human combustion (SHC) as both most amazing and most terrifying. I have had no direct contact with SHC nor any firsthand witnesses of it, so the evidence for it may be questioned. Yet it is not expected to be a "psychic" phenomenon associated with the psyche (it is hard to conceive of someone "willing" themselves to burst into flames). It is presently scientifically inexplicable, and, if true, a remarkable phenomenon.

Related to SHC are the psychic burns also described earlier. Again, I have had no direct contact with this phenomenon, but I have discussed it with certain observers who have experienced it in person. Here, however, the accounts of subjects claiming feelings of extreme heat when undergoing psychic healing (Chapter 4) are relevant; at no time did this actually occur, according to my measurements. It was all in the mind! Could psychic burns be similar, yet so strongly experienced psychologically as to cause the burns to appear by sheer willpower?

Psychic healing itself presents a scientific/medical problem. We have suggested that it may well be possible to explain psychic healing as brought about by the psychological influence of the healer on the patient. This feature of the healing process is known to a certain extent in the medical profession. Some surgeons are well known to cause incisions which will heal more rapidly than those produced by their less skilled colleagues. It has been ascribed to "bedside manner," which is another term for this important psychological relation. Pain can be reduced by suitable orientation of the patient, as already noted in the use of acupuncture in surgery, or in psychological counseling of patients.

We have not given a scientific explanation of pain alleviation or accelerated wound healing as achieved by certain psychic healers. But such features may well be effected by the patient's own bodily control system. This latter would very likely also achieve spontaneous remission of cancer or other fatal diseases. The answer thus lies in the way the mind controls the body's mechanism for resistance to disease, a problem at the forefront of medicine. There is presently a great deal of interest in the way the control of the mind over bodily health can be increased. Groups of terminal cancer patients are being taught meditation practices in an attempt to increase resistance to the spread of their disease. There are biofeedback methods to help the patient learn to reduce his or her levels of tension, as we mentioned earlier. Usually this is done by increasing the skin resistance of the palms of the hands. Less frequently, brain-wave patterns are monitored so the subject learns to produce a lower and more synchronous brain activity. These techniques seem of value, especially in the reduction of pain, though have not yet been fully assessed. There are also successes in the control of hypertension and epilepsy by such biofeedback methods, and no doubt this control will spread to other diseases.

A further question of some scientific interest is the manner in which some faith healers appear to have uncanny powers of diagnosis. It would seem correct to call such healers "sensitives." But what visual cues are they picking up from their patients? I suggest that medical science might usefully learn from a careful study of certain healers what constitutes their successful "bedside manner," both from the point of view of diagnosis and of helping the patient to help himself.

There are somewhat different problems in connection with the various apparently successful tests of GESP (telepathy and clairvoyance) and precognition, described in Chapters 5, 6 and 7. Their apparent success was ascribed to various possible faults in the experiments, but the details of such faults were not given explicitly. In order to find these faults one has to repeat the reported tests oneself—a "do-it-yourself" program.

A good portion of this book is a record of my results on carrying out such a program into various paranormal phenomena. Everything that I investigated turned out either to have a scientific explanation, such as the electrical explanation of certain psychokinetic results, or did not occur at all under careful test conditions. The earlier results of others in these latter cases were found to be explicable under the headings of mischief, fraud, credulity, fantasy, memory, cues and fear of death. With this evidence and the fact that all of these phenomena disagreed completely with scientific results, we have to conclude that the paranormal has "disappeared."

As we said at the beginning of the chapter, there are a few psychic phenomena which are still difficult to explain completely and finally, these being spontaneous human combustion and psychic burns and cures. It has already been pointed out that their character is almost certainly not paranormal. Faith healing is ultimately to be explained in medical terms. SHC will either be a complete fabrication or arise from some natural cause, such as a spark caused by static electricity, igniting highly inflammable gases from the subject's insides.

The supernatural has thus become completely natural. *The paranormal is now totally normal. ESP is dead.* Such disappearance of the supernatural is inevitable if we weigh it against science. I started my investigation with an open mind; the scales were not loaded on behalf of science. On the evidence presented in this book, science has won.

Yet it may never do so in the minds of many. There will always be human gullibility. Five million people will buy a book based on incorrect evidence. As many will buy a book in which it is claimed that plants feel emotions, a book full of fantasies masquerading as science. Millions buy books on pyramid power and other such nonsense, again dressed up in a scientific guise. There are even mature scientists who can seriously investigate (using techniques of the utmost questionableness) phenomena which their scientific education should indicate are impossible. I have done so myself. How can we be so blind?

I think we can discern various reasons for this. Firstly, it is relatively easy for the man in the street to comprehend writings of a pseudoscientific nature about the supernatural. This is because such books do not require any knowledge of the vast body of scientific thought in order to understand them. Truly scientific works, on the other hand, sadly often fall short of comprehension for many potential readers.

Compare the millions of readers of *The Bermuda Triangle* or *Chariot of the Gods* with the mere thousands who have read and understood P. A. M. Dirac's masterpiece *Quantum Mechanics.* Yet the latter is one of the great books of our century, in spite of not having been heard of, let alone read, by the great majority. Nor is its author's name known to the vast majority; yet Dirac is one of the greatest scientists of this century, with the construction of quantum field theory and the theory of the electron and positron to his credit.

A second reason is the great interest of the man in the street in powers which he thinks he may himself possess and be able to develop quite rapidly. Such magical powers have always appealed to those with otherwise limited opportunities for self-advancement.

There are also the situations we have encountered in our own search where fraud is clearly absent, especially that of faith healing. Small wonder that the faith of the believer is reinforced when true healing does occur.

Yet it is difficult to make sense of the vast wealth of reports on supernatural phenomena. They range from secondhand (or worse) fragments, which may have been considerably distorted by the time they reach you, to what are purported to be precise, scientifically controlled tests irrefutably showing that this or that paranormal power does exist. Such reports have even appeared, though very infrequently, in the established scientific press. Results of some telepathy tests at Stanford have been published in England in the "prestigious" scientific establishment journal *Nature* and also in the reputable journal of the American Institute of Electrical and Electronic Engineers. However, the editorial comments published simultaneously with the telepathy article in *Nature* indicated that the paper did not really reach the standards requisite for publication, but was published so that other scientists might know of what the Stanford research consisted.

A twenty-three-page paper has also been published in the highly respectable *Journal of Nervous and Mental Disease,* documenting sixteen hundred cases of people who claim to remember specific details of earlier lives. Thus the seal of academic approval through publication in a reputable journal appears to have been placed on reincarnation. Yet here again, the editor of the journal is unhappy about the "reincarnation" interpretation of the cases presented.

Such publication cannot be used to argue one way or the other for the validity of ESP findings. It should be said quite bluntly that a not inconsiderable proportion of papers published in so-called scientific journals contain errors of one sort or another, some of them major ones. The publication of a handful of papers on the paranormal, especially in such a hesitant and apologetic manner, cannot be used in any way to suggest that the scientific establishment is weakening its opposition to the supernatural.

Nor do the many debates over cases published in books or in

the "parapsychological" literature help to clarify the position. Even among psychical researchers there is little unanimity as to what tests have proved completely convincing. An example of this is the assessments given by two groups of authors on published tests of telepathy. Out of 142 such publications only six were judged impeccable by the first group of five authors, while a group of two other authors only agreed unhesitatingly on one of those six. There is no unity of opinion even among leading psychical researchers as to what constitutes valid evidence.

The case for the paranormal is certainly not helped by the often absurd claims made by some of the more vociferous believers in support of their faith. The list of distortions and half-truths used to bolster the cause pours out in an apparently unending stream of popular and semipopular books on all areas of the supernatural. One of the most nonsensical of these is that physics has discovered such odd things that yet another one "won't hurt." Brian Inglis, a writer generally sympathetic to the supernatural cause (though not in an earlier quotation from him), wrote recently, "Research in nuclear physics has uncovered such a collection of crazy forces that the addition of another one would now hardly surprise them." This line has been used by, among other well-known writers, Arthur Koestler and Lyall Watson. Yet few nuclear physicists would accept such remarks other than for the untutored errors which they are. No wonder scientists are repulsed by the supernatural if such distortions of modern science are propounded by its supporters.

What can the man in the street conclude with so much disagreement? It is not surprising that he turns to where his fantasy leads him—*Chariots of the Gods, The Bermuda Triangle,* pyramid power and all. He or she will be searching for an answer to the unspoken question as to why they are here and what it all means. But it has to be a loud and clear answer and one that is short and sharp. That is presently absent from the more serious literature.

The supernatural would seem to give such an answer, either from the spirits of those who have "passed on," from the mouths of alien beings, from those who have lived many times or from "psychics" who are prepared to jump in where angels fear to tread.

The answer I give here now is short and sharp: the mystery of existence is not to be gained by searching for strange paranormal powers possessed by humans. It is to be gained by looking more closely at the beautiful edifice of science, to see how the whole of existence—both of ourselves (all living beings) and of the material world—is to be understood in a unified manner. We, as humans, are at one with the rest of existence. The basis of all is energy, in its various manifestations. The question now to be answered is why those manifestations of energy are there in the first place.

We still have to turn to ourselves as manifestations of energy, and in particular to consider the part played by our minds in all this. However much we investigate and develop the control of the mind over the body, we still have to understand what this entity called "mind" really is. Indeed, we might regard the problem of mind as the crux of the matter, since it is by the mind that paranormal phenomena are supposed to occur. Without mind being present, we do not expect matter suddenly to disappear or perform other spiritualist effects. It is rather unlikely, for example, that coats or hats have a spirit associated with them.

It is, in fact, the problem of mind that brought me into contact with the supernatural in the first place. I felt it important, as a scientist, to investigate these latter phenomena as having great relevance to the mind/body problem.

The problem of mind has been investigated from many viewpoints. We are clearly restricted here to discussing it from the materialistic angle. The question we have to answer is how the mind can be constructed from purely material entities. There have been many attacks on such a materialistic program and some would claim that it will always fail. I can only reiterate here that if it does, then we cannot usefully discuss the mind further: it will not be objectively describable. Again, we will assume that such a gloomy prognostication is not true, and start to give a physical theory of mind. We certainly will not dismiss the mind as nonexistent, as some materialists are supposed to do. It is undoubted that the mind has played a crucial role in evolution, especially in the emergence of man as the dominant animal.

We turn to the quest for a physical theory of mind. The requirement of such a theory is that it construct what is apparently a nonphysical entity—mind—from a purely physical one—matter. The only reasonable theory that can hope to do this is what I would call the "relational theory" of mind. In this approach the mind is regarded as the set of relations among physical events such as firings of nerve cells.

It is possible to give an analogy to this "relational" theory of mind. If we consider the numbers, 1, 2, 3, etc., we can construct the relations between them, such as $1 < 2, < 3 < \ldots$, and so on. These relations are not themselves numbers, but are clearly features which are closely related to them. In the same way, the relations between firing patterns of nerve cells in the brain at two different times are not themselves physical quantities. These relations are, however, of great relevance to activity. In particular, close correspondence between brain activity at one time and at an earlier one is clearly of great relevance. It may be of importance to be able to relate similar brain activities, especially if the results of the earlier activity were traumatic to the subject.

We thus see that the brain is expected to be so constructed as to allow past brain activity to be utilized rapidly in assessing future courses of action. The brain should therefore be a "relational machine," adept at relating present brain activity to relevant past action. The mind then develops as the complex set of relations of a given brain activity with all past actions.

A child will be born with almost no past brain processes to utilize, so will have a very "weak" mind. As he or she grows older, the ever-increasing amount of past brain activity, assumed to be stored in a suitable fashion, will color present activity ever deeper. Ultimately the adult mind will be formed, though since new experience is always occurring no person's mind should ever cease increasing in complexity.

The mind is thus considered as the set of relations between ongoing and past brain activity. The mind is not physical, but can be utilized to affect physical activity. For example, certain crucial past relations might be found to be highly relevant to present brain activity. The set of relations and the underlying nerve cells are inextricably bound up together, exactly as numbers and their relations are. Without one, the other would not exist.

According to this model, the mind would be powerless to act, or even exist, without the brain. Disembodied minds are impossible, as is any power of mind over body, other than that mediated through the nerve cells controlling hormone levels, muscle activity and other biochemical features of body activity. Any explanation of the paranormal is thus to be found through the action of the body. As we saw earlier, that can only be by means of electromagnetic action of some form or other. It might be by static electricity or by radio-wave emission that this is achieved. But we come back again to the electromagnetic spectrum, suitably utilized by the body, to achieve paranormal effects. We have already investigated that and found, on both theoretical and experimental grounds, that there is no abnormal electromagnetic signal during paranormal events.

Our discussion of the mind has, in fact, led us to a new avenue to explore—that of the nature of consciousness itself. Sir John Eccles, Nobel Laureate in Physiology in 1963, spoke along such lines in his address at the banquet of the Parapsychology Convention held in Utrecht in 1976. "The most paranormal thing of all," he said, "is how I can move my finger when I so will it. The mind is *the* problem to explain in any parapsychological investigation." I can add to that a conclusion we can draw from our search for the supernatural. The mind is the only paranormal thing left to understand. That is indeed a challenge to take up.

Here then, we must stop, on the frontier of new country. A simple physical map of the mind has been drawn, though containing few details. Only the bare outlines are apparent, and they may prove a hallucination. Even so, they certainly represent something definite since, as Descartes said, "I think, therefore I am." There will be no elusive phenomenon to track down in such a search. It is all there, sitting under our noses, or more appropriately, inside our skulls. Only the unraveling of the human mind will, I think, enable us finally to lay to rest the spirit of the supernatural.

Further Reading

BATCHELDOR, K. J. "Data Tape-Recorded Experimental PK Phenomena." *Journ. Soc. Psych. Res.* 47 (1973) : 69–89.

BROWN, M. H. *PK: A Report on Psychokinesis.* New York: Steinbooks, 1976.

COX, W. E. "Precognition: An Analysis I and II." *Journ. Soc. Psych. Res.* 50 (1956) : 47–58 and 97–107.

DIRAC, P. A. M. *The Principles of Quantum Mechanics.* Oxford: Clavendon Press, 1930.

EBON, E. (ed.) *The Amazing Uri Geller.* New York: Signet Books, 1975.

EISENBERG, H. *Inner Spaces.* Ontario: Musson Book Co., 1977.

EVANS, C. *Cults of Unreason.* London: Harrap, 1973.

FODOR, N. *Encyclopaedia of Psychic Sciences.* Secaucus: University Books, 1966.

GELLER, U. *My Story.* New York: Praeger, 1975.

GLANVIL, J. *Saddicimus Trumphotus or A Full and Plain Evidence Concerning Witches and Apparitions.* In 2 parts, 4 ed. London: 1721.

GOULD, A. *The Founders of Psychical Research.* London: Routledge & Kegan Paul, 1968.

HANSEL, C. E. M. *ESP—A Scientific Evaluation.* New York: Scribner, 1966.

HARMON, H. B. *Edgar Cayce.* New York: Paperback Library, 1970.

HERBERT, B., KEIL, H. H., PRATT, J. G., and ULLMAN, M. "Directly Observable Voluntary PK Effects." *Proc. Soc. Psych. Res.* 56 (1976) : 199–235.

KOESTLER, A. *The Roots of Coincidence.* London: Hutchinson, 1972.

KOGAN, I. M. "Is Telepathy Possible?" *Telecom. Radio Eng.* 21 (1966): 75–81.

KRIPPNER, S., and RUBIN, D. (eds.) *The Energies of Consciousness.* New York: Gordon & Breach, 1973.

MITCHELL, E.D. (ed.) *Psychic Explorations.* New York: Putnam's, 1974.

MISHLOVE, J. *The Roots of Consciousness.* New York: Random House, 1975.

MOODY, R. A., JR. *Life After Death.* Atlanta: Mockingbird Books, 1975.

MUSES, C., and YOUNG, A. M. (eds.) *Consciousness and Reality.* New York: Avon Books, 1972.

MYERS, F. *Human Personality and Its Survival of Bodily Death.* Toronto: Longmans Green, 1954.

OWEN, A. R. G. *Can We Explain the Poltergeist?* New York: Garrett Pub., 1964.

————. *Psychic Mysteries of Canada.* Toronto, Montreal, Winnipeg, Vancouver: Fitzhenry and Whiteside, 1975.

OWEN, I. M. *Conjuring Up Phillip.* Toronto, Montreal, Winnipeg, Vancouver: Fitzhenry and Whiteside, 1976.

PANATI, CH. (ed.) *The Geller Papers.* Boston: Houghton Mifflin, 1976.

PEARSALL, R. *The Table Rappers.* London: Book Club Associates, 1973.

PUHARICH, A. *Beyond Telepathy.* London: Souvenir Press, 1962.

————. *Uri.* New York: Anchor Press, 1974.

PUTHOFF, H. E., and TARG, R. E. "A Perceptual Channel for Information Transfer Over Kilometer Distances: Historical Perspectives and Recent Research." *Proc. Inst. Elec. and Elects. Eng.* 64 (1976) : 329–354.

————. *Mind-Reach.* London: Granada, 1978.

RANDALL, J. "Recent Experiments in Animal Parapsychology." *Journ. Soc. Psych. Res.* 46 (1972) : 124–135.

RANDI, J. *The Magic of Uri Geller.* New York: Ballantine Books, 1975.

RHINE, J. B. *The Reach of the Mind.* London: Faber & Faber Ltd., 1948.

RHINE, J. B., PRATT, J. G., SMITH, B., STUART, C., and GREENWOOD, J. *Extra-Sensory Perception After Sixty Years*. Boston: Bruce Humphries, 1966.

RHINE, L. E. *Hidden Channels of the Mind*. New York: Sloane, 1961.

———. "Psychological Processes in ESP Experiences, I. Waking Experiences. II. Dreams." *Journ. Parapsych.* 26 (1962) : 88–111 and 171–199.

———. *ESP in Life and Lab: Tracing Hidden Channels*. New York: Macmillan, 1967.

RIESS, B. F. "A Case of High Scores in Card Guessing at a Distance." *Journ. Parapsych.* 1 (1937) : 260–263.

ROLL, W. G. *The Poltergeist*. New York: Signet Books, 1972.

ROLL, W. G., and PRATT, J. G. "The Miami Disturbances." *Journ. Amer. Soc. Psych. Res.* 65 (1971) : 409–454.

RYALL, E. *Second Time Round*. London: Neville Spearman, 1974.

SAGAN, C., and PAGE, T. (eds.) *UFOs: A Scientific Debate*. Ithaca and London: Cornell Univ. Press, 1972.

SCHMIDT, H. *Journ. Parapsych.* 33 (1969) : 99–108.

———. "Mental Influence on Random Events." *New Scientist* 50 (1971) : 757–758.

SPINETTI, E. "The Effect of Chronological Age on GESP Ability." Presented at the Parapsychology Convention, Utrecht, 1976.

TABORI, P. *Beyond the Senses*. London: Souvenir Press, 1971.

TAYLOR, J. G. *Superminds*. London: Macmillan, 1973.

TAYLOR, J. G., and BALANOVSKI, E. "Can Electromagnetism Explain ESP?" *Nature* 275 (1978) : 64.

———. "Is There Any Scientific Explanation of the Paranormal?" *Nature* 279 (1979) : 631–633.

———. "A Critical Review of Explanations of the Paranormal." *Psychoenergetic Systems* 1 (1979).

———. "A Search for the Electromagnetic Concomitants of ESP." *Psychoenergetic Systems* 1 (1979).

TAYLOR, J. G., BALANOVSKI, E., and IBRAHIM, R. "The Interpretation of Effects Seen in Kirlian Photography." *Psychoenergetic Systems* 1 (1979).

TOYNBEE, A., et al. (ed.) *Man's Concern with Death*. London: Hodder & Stoughton, 1968.

VASILIEV, L. L. *Experiments in Distant Influence.* London: Wildwood House, 1976.

WATSON, L. *Supernature.* London: Hodder & Stoughton, 1973.

WHITE, R. A. (ed.) *Surveys in Parapsychology.* New Jersey: Scarecrow Press, 1976.

WILHELM, J. L. *The Search for Superman.* New York: Simon and Schuster, 1976.

WOLSTENHOLME, G. E. W., and MILLAR, E. C. P. (eds.) *Extrasensory Perception.* London: J. & A. Churchill, 1956.

Index

accident-avoidance, 150–151
acupuncture, 34, 41, 163
Adam, Herr, 18–19
altered states of consciousness, 67
alternate science, 22
Anderson, Carl David, 82
animals, haunted houses and, 136
Annemarie (psychic heater), 12
Arigo (psychic surgeon), 32, 35–36, 41
assassinations, probability of, 74, 150
auras, human, 3. *See also* Kirlian photography
automatic writing, 137

Balanovski, Eduardo, 8, 37, 41, 43–44, 71–72, 83, 103–104, 121, 132, 138
Balinese, telepathy among, 16, 59
ball lightning, 157
Batcheldor, K. J., 101
belief, need for, 108, 161
Berlitz, Charles, 159
Bermuda Triangle, The (Inglis), 159, 165, 167
Bermuda triangle phenomenon, 159
biocommunication, 57
biofeedback, 37–38, 163–164
bioplasma body, 43
black-body radiation, 29, 40, 54–55, 158
Bleak House (Dickens), 10–11
Bloxham, Arnall, 125–126, 130, 132
Boer, W. de, 71
Bohr, Niels, 73
Brady, Ian, 144
brain activity, mind and, 168–170
Brazil, psychic healers in, 32, 33–36
British Society for Psychical Research (SPR), 134, 146
Brookes-Smith, Colin, 101
bubble chambers, 47–48
Buckhout, Robert, 107–108
Burr, H. S., 41
Bushnell, Henry, 15–16

Cade, Max, 132
Capital Radio reincarnation programs, 130–132
Carter, Jimmy, 3
causality, 23–24

dispersion relation and, 81–82
violations of, 80–83
Cayce, Edgar, 14
Cazzamali, F., 64
Chariot of the Gods, 165, 167
cheating, by mediums, 109–110
China, dowsing in, 140
Christianity, immortality of soul in, 152
Churchill, Sir Winston, 148
clairvoyance, 46–56, 146, 164
cases of, 14–16, 46–48
in crime solving, 108, 143–145
defined, 1
energy source for, 21, 52–56
investigation areas of, 51–52
radio wave transmission and, 52–55, 79–80
telepathy and, 58–59, 60
tests for, 48–56, 58–59, 60, 69
Clark, Bill, 12
Clarke, Arthur, 117
Coeur, Jacques, 125
coincidence, 84, 108, 147–148
in accident-avoidance, 150–151
in fantasy, 145–146
combustible gas, in spontaneous human combustion, 22
Committee for the Scientific Investigation of Claims of the Paranormal, The, 4
Condon, Edward, 3, 111
consciousness, materialistic models of, 26–27
corona discharge, 44
corona discharge photography. *See* Kirlian photography
cosmic ray tracks, 82
Cox, Sergeant W., 98
Cox, W. E., 16–17, 150–151
crime solving, clairvoyance in, 108, 143–145
Croiset, Gerard, 108, 143–145
cryptomnesia, 127

dead, communication with, 2. *See also* mediums
Dean, E. Douglas, 75

death:
    experiences of, 152–154
    fear of, disbelief and, 152
    religion and, 151–152
Defense Department, U.S., 3–4
déjà vu, defined, 2
demons, 126–127
Descartes, René, 170
description, in scientific investigation, 24–25
diathermy, 39
dice test, for psychokinesis, 100–102
Dickens, Charles, 10–11
dielectric resonance, 21
Dimbleby, David, 5
Dirac, Paul A. M., 82, 165
dispersion relation, causality and, 81–82
distant healing, 13, 33, 36–37. *See also* healers; healing; meditation; psychic healing
Dixon, Jeane, 74, 150
door banging, 86
door opening, 86
Downey, Lesley Ann, 143–144
dowsing, 1, 83, 146
    clues in, 143, 146
    history of, 140
    magnetic field sensing theory of, 69–72
    by maps, 142
    medical diagnosis and, 140–141
    process of, 139–140
    tests of, 141–143
    unconscious muscular reaction and, 142–143
    well comparisons and, 141
dreams:
    birth trauma and, 150
    nightmares as, 150
    precognition in, 73, 74–76
    premonition in, 149–150
    in telepathy experiments, 66–67
Duke University clairvoyance tests, 48–50, 53
Durant, W. C., 159
dynamism description test, 160

Eccles, Sir John, 170
Edivaldo, 34–35
Edwards, Happy, 24, 32
eidetic memory, 133
electrical sensations, in healing, 12–13, 39
electromagnetism, 27–30
    fifth force and, 28
    measurement of, during healing, 39–42
    solitons in, 157–158
electromyograph (EMG), 119–120
electrons, 82–83

energy:
    bioplasma, 157
    in clairvoyance and telepathy, 21
    degrees of, 20
    in heat production and healing, 20, 32, 39
    measurement of, during psychic healing, 39–45
    in psychokinesis and poltergeists, 20–21
    sensitivity to transmission of, 21
    unknown source of, in psychic phenomena, 20–21
ESP (extra-sensory perception), 2
    altered states of consciousness and, 67
    belief in, among businessmen, 160–161
    in children, 62–63
    discounted, 164–170
    research in U.S.S.R. on, 3–4
    telepathy and clairvoyance as (GESP), 58, 62–63. *See also* psychic phenomena
ESP cards, 48–49
etheric body, 43, 154
expectation, in paranormal events, 108
Extra-Sensory Perception After Sixty Years (Rhine), 48
extraterrestrial life, 159

faith healing, 166
    in Bible, 31
    defined, 1–2. *See also* healers; healing; psychic healing
fantasy, 108, 129–130
    coincidence and, 145–146
    about life after death, 151–153
Faraday, Michael, 101, 122
Feynman, Richard P., 82
Fletcher, John, 127–130
"Forty Demons Slaughter" case, 126–127
Fox, Catherine, 96–98, 110–111, 112–113
Fox, John, 96–97
Fox, Leah, 97–98, 111, 112–113
Fox, Margaretta, 96–98, 110–111, 112–113
France, UFO investigations in, 3
fraud:
    detection of, 116–124
    in Geller case, 114
    mechanical force and, 115–124
    among mediums, 109–110
    in psychokinesis, 110–111, 114–124
    substitution and, 115–116
    in UFO photographs, 111
Geller, Uri, 4–8, 122, 124, 161
    conjuring ascribed to, 114, 117–118

spoon bending by, 5–9, 28, 110, 113–118
GESP (generalized extra-sensory perception), 58, 62–63
ghosts, 127, 134–137
ghost telephone calls, 18–19
Giorgini, Maria, 12–13
Glanvil, Joseph, 85–86
gravity, 27, 56
Gurney, Edmund, 134

hallucinations, 27, 108, 135
Hand of Fate, The, 147
Hansel, C. E. M., 49–50, 61
Harvalik, Zaboj, 70–72
Hasted, John, 105
haunted houses, 134–135, 146
  animals and, 136
  investigations of, 134, 138–139
healers:
  diagnosis by, 164
  distant healing by, 13, 33, 36–37
  laying on of hands by, 12, 32, 36
  meditation with patients of, 13, 33, 36–37
  patient relationship to, 37, 45, 163
  psychic surgery by, 32–36
  spiritual guidance of, 31–32
  trances of, 12. See also faith healing; healing; psychic healing
healing:
  biofeedback and, 37–38, 163–164
  environment and, 32
  placebo effect and, 38–39
  psychological factors in, 31, 32, 36–39, 45, 163
  spontaneous, 31, 32
  will to live and, 38–39. See also faith healing; healers; psychic healing
heat, psychic production of. See psychic heating
Henry, Mr., 109
Herbert, B., 94
Herrmann poltergeist case, 91–92
Hindley, Myra, 144
History of England (Macaulay), 128
Home, D. D., 88
hypnosis, 34
  cryptomnesia and, 127
  in fraud, 116
  precognition and, 76–77
  regression during, as therapeutic, 131
  reincarnation and, 125–127, 130–132, 146
  subconscious during, 48
  suppressed memory and, 133–134

Ibrahim, Ray, 8, 44
immortality, 152, 154–158
infrared radiation, 29
Inglis, Brian, 159, 167

intuition, in business, 159–161
ion-acoustic plasma waves, 157
Islam, immortality of soul in, 152

Jesus, 31, 73
Joseph (biblical), 73
Journal of Nervous and Mental Disease, 166
Julie K., straw rotation by, 105

Keil, H., 94
Kennedy, John F., 62, 74
key bending, 122–124
Kirlian, Semyon, 43
Kirlian photography, 3–4, 39, 43–44, 157
kitomine, 154
Koestler, Arthur, 167
Kogan, I. M., 68, 79
Kübler-Ross, Dr., 153
Kulagina, Nina, 94–95, 105

Lambert, W. G., 88
laying on of hands, by healers, 12, 32, 36
Leek, Sybil, 13, 37
levitation game, 121–122
Lewis, Bill, 140–141
life after death, 24, 151–158, 166–167
  descriptions of, 152–153
  etheric body and, 154
  explanations for visions of, 154
  loved ones and, 152–153
  as physical energy continuation, 154
  proof of, 155–158. See also reincarnation; soul
Lourdes, cures at, 14
LSD, 153
Luther, Martin, 140

McAdam, Pat, 144–145
Magic of Uri Geller, The (Randi), 117
magnetic field sensors, human, 69–72
Magnusson, Magnus, 132
Maigret, Pamela Painter de, 17–18, 19
Maimonides Medical Center, dream experiments at, 66–67, 75–76
Manning, Matthew, 83
Martin, Dorothy, 50, 60
Martin-Stribic tests, 50, 60
materialism:
  consciousness and, 26–27
  reductionist program of, 27–28
  in scientific investigations, 25–27
meditation:
  in biofeedback, 37–38, 163
  in distant healing, 13, 33, 36–37
mediums, 109–111
  automatic writing by, 137
  cheating by, 109–110
  clues for, 137–138
  direct voice, 2, 137
  telepathy and, 137

Mellon, Mrs., 109
memory, suppressed, 108, 132–134
mental experiences, non-quantitative, 25–27
Meredith, Reverend, 129
Mershon, Mrs. K. E., 16
microwave emission, human, 29–30
  measurement during psychic phenomena of, 29–30
  in poltergeist phenomena, 29–30
microwave radiometry, 29
mind/body problem, 168–170
Mishlove, Jeffrey, 75
Mitchell, Ed, 58, 63
Moana (healer), 13, 37
Moody, R. A., 152–153
Moorgate tube disaster, 74–75, 150
Moors murder case, 143–144
Moses, Stainton, 98
Moss, Thelma, 43, 62, 67
Motoyama, Dr., 41, 42
muscular action, unconscious, 101, 121–122
  in dowsing, 142–143
Myers, Frederick W. H., 57, 142
Myers, W., 136

Nature, 166
negative-energy particles, 82–83
neutrinos, 47–48
neutron, clairvoyant description of, 46–48
nightmares, 74, 150
noisy ghosts. See poltergeists
nuclear force, 27, 56

oracles, 73
Osis, Karlis, 63
out-of-body experiences, 152–154

pain alleviation, by psychic healers, 163–164
"para," as prefix, 1
paranormal, defined, 1
Patel, Chandra, 37–38
Pearce, H., 49–50, 60
Pearsall, Ronald, 107
Perrin, Serge, 13–14
personality transfer, 154–155
Phantasms of the Living (Gurney), 134
Philip group table case, 98–100, 101
physics, causality violations in, 81–83
picture guessing, in telepathy, 63–64, 114
Piper, Leonora, 137
Pitt-Rivers, General, 107
placebo effect, 38–39
Playfair, Guy Lyon, 34–35
Podmore, Frank, 136–137
poltergeists, 23
  cases of, 18–19, 85–95, 96–100, 107

defined, 18
energy source for, 20–21
microwave emission and, 29–30
noises by, 85, 88–92, 96–100
objects moved by, 86–94
phenomena classification of, 86
underground water and, 88, 90
positrons, 82–83
possession, 127
Pratt, J. G., 49–50, 60, 91, 93–94
Pratt-Woodruff tests, 50, 60
precognition, 23–24, 73–84, 164
  in Bible, 73
  in business, 159–161
  cases of, 16–17, 74–75
  as causality violation, 80–83
  defined, 1, 16, 74
  in dreams, 73, 74–76
  energy movement during, 21
  high energy physics and, 81–83
  hypnosis and, 76–77
  vs. prediction, 73–74, 80
  tests for, 77–80, 83
prediction, 73–74, 80
premonitions, 148–151
  in dreams, 149–150
Press, Frank, 3
Price, Pat, 53, 69
Proceedings of the British Society for Psychical Research, 61–62
prophets, 73
psychic healing, 31–45
  bleeding and, 12–13, 34–35
  cases of, 12–14, 33–36
  closure of incisions following, 34–36, 163
  defined, 1–2
  documentation of, 32–33
  electromagnetic theory of, 39–42
  energy measurement during, 39–45
  energy source of, 20, 32
  illusions in, 34–35
  infection and, 36
  patients' sensations during, 12–13, 34–35, 39
  psychic surgery in, 32–36
  unexplained aspects of, 163–164
  validity of, 32–33. See also faith healing; healers; healing
psychic heating, 12, 20
  burns caused by, 20, 163
  in healing sessions, 32, 39, 42–43
  poltergeist activity and, 86, 94–95
  in straw rotation, 105–106
  as unresolved phenomenon, 163. See also spontaneous human combustion
psychic phenomena:
  antagonism toward scientific analysis of, 7, 24

conclusions about, 146, 161, 165–170
electromagnetic explanation of, 28–30
government interest in, 3–4
role of mind in, 168–170
Russian studies of, 3–4, 63, 64–66, 68, 94–95, 102–103, 157
scepticism toward, 4, 7
scientific investigation of, 6–9, 28–29
unresolved problems in, 162–165
widespread acceptance of, 2–3, 7, 165–168
psychic surgery, 32–36
psychokinesis (PK), 85–108, 164
cases of, 17–18, 94–96, 102–107
defined, 1, 17
dice test for, 100–102
energy source of, 20
fraud in, 110–111, 114–124
heat-caused, 106–107
poltergeists and, 18
static electricity and, 102–105, 107
psychometry, by dowsers, 140
Puharich, Andrija, 35–36, 46
Puthoff, H. E., 52, 54, 78–79, 81–82, 161

quantification, in scientific investigations, 24–25, 27
Quantum Mechanics (Dirac), 165

radioactivity, 27, 56
radio waves, 39–40
causality-satisfying, 79–80
in clairvoyance, 52–55, 79–80
retarded, 79–80
soul as manifestation of, 157–158
in telepathy, 23, 64–66, 68–69
wavelengths of, 29
rainmaking, 150
Randi, James ("the Amazing"), 103–104, 114, 117, 118, 161
random number generator, 77–78
rapping:
as poltergeist activity, 85, 96–100
during psychic sessions, 30
Reach of the Mind, The (Rhine), 100–101
reductionist program of materialism, 27–28
constituents of objects in, 27
forces in, 27–28
Reeser, Mrs., 11
reincarnation, 125–132, 135–136, 146, 166
as transfer of personality, 154–155. See also life after death
Rhine, Herbert, 149
Rhine, J. B., 48–49, 51, 53, 60, 149
psychokinesis tests of, 100–101

Rhine, Louisa E., 60, 149
Riess, B. F., 60–61
Rocard, Y., 70
Roll, W. G., 91, 93–94
Roots of Consciousness, The (Mishlove), 75
Rosenheim poltergeist case, 18–19, 85, 94, 107
Russia. See U.S.S.R.
Ryall, Edward, 127–130, 133, 134, 146

Sauchie poltergeist case, 88–91
Schmidt, Helmut, 77, 100, 102
science:
description in, 24–25
explanations of paranormal phenomena and, 22–24
materialistic investigations in, 25–27
mental experiences and, 25–27
paranormal contradictions of, 23–24, 27
quantification in, 24–25, 27
Scott-Russell, soliton observation by, 156
Second Time Around (Ryall), 127–128
Seifer, Marc, 122–123
seiketsu points, 41
sensitives, 21, 46
sensory-deprivation, 153–154
Shtrang, Shimson "Shipi," 114
Sidgwick, Professor, 134–135, 137
simultaneity, 155
sleeping, in telepathy experiments, 65–66
Sloan, Alfred P., 159
Smith, Anthony, 8
Smubler, Howard, 122–123
solitons, 156–158
soul:
as etheric body, 154
immortality of, 152, 154–158
proving existence of, 155–158
in psychic occurrences, 1
as radio-wave manifestation, 157–158
as soliton, 156–158
Spinetti, Ernest, 63
spirits, healing guidance by, 31–32
spontaneous human combustion (SHC), 10–12, 20–22, 164–165
combustible gas theory of, 22, 165
dielectric resonance theory of, 21
difficulty of studies on, 21–22
microwave emission during, 30
as unresolved phenomenon, 162–163
spoon bending, 5–7, 8–9, 23, 28, 83
as epidemic, 111–115
fracture analysis and, 120–121
fraud in, 110, 114–124

spoon bending (*continued*)
  by Geller, 5–9, 28, 110, 113–118
  letter balance method for testing, 118–119
Stanbridge, George Henry, 145, 147, 151
Stanford Research Institute (SRI), 161
  clairvoyance tests at, 52–56, 58–59, 69
  precognition tests at, 78–80, 83
static electricity:
  health and, 40–41
  psychokinesis and, 102–105, 107
statistical analysis, 49–51
Stevenson, Ian, 60, 127–128
stigmata, 12–13
straw rotation, 105–106
Stribic, Frances, 50, 60
subconscious knowledge, during trance states, 48
Superminds (Taylor), 118
supernatural, defined, 1
supraphysical bodies, 135–136

table moving, 30
  Faraday tests of, 101, 122
  Philip group and, 98–100, 101
"Talk In" show, Geller appearance on, 5–6
Targ, R., 52, 56, 78–79, 81–82, 161
Taylor, Michael, 126–127
telepathy, 57–72, 164
  among Balinese, 16, 59
  cases of, 16, 57–61
  clairvoyance and, 58–59, 60
  defined, 1, 16, 57
  distance and, 61, 63–64
  dream experiments and, 66–67
  emotional content of messages and, 60, 61–62, 67
  energy source for, 21, 67
  fraud in, 111
  Geller's power of, 5
  by mediums, 137
  men and, 67
  in primitive communities, 59

radio wave theory of, 23, 64–66, 68–69
  Russian experiments in, 63, 64–66, 68
  sleep experiments and, 65–66
  spontaneous, 59–60, 61–62, 67
  in women, 60
Tizane, E., 86
Todaro, Teodore, 13
Toth, Robert, 4
trance states:
  former lives revealed in, 125–126
  subconscious knowledge during, 48.
  See also hypnosis
Ullman, M., 94
unidentified flying objects (UFOs), 3, 111, 158–159
U.S.S.R.:
  bioplasma energy studies in, 157
  ESP research in, 3–4
  psychokinesis in, 94–95, 102–103
  telepathy experiments in, 63, 64–66, 68

Vasiliev, L. L., 57–58, 64–66
Vinogradova, Adamenko, 96, 103
Vinogradova, Alla, 17–18, 95–96, 102–103, 105

Wallace, George, 149
watch starting, 5–6, 114
water, structural modification of, 43
Watkins, Graham, 136
Watson, Lyall, 5, 167
Weatherhead, Leslie, 155
weeping of blood, 12–13
white light, experienced by psychics, 32
Willie G., straw rotation by, 105
Wilmot, Mr., 14–15
Without a Trace (Berlitz), 159
Woodruff, Mr., clairvoyance tests by, 50

Youatt, Captain, 15–16

Zeca (healer), 33–34